SpringerBriefs in Astronomy

More information about this series at http://www.springer.com/series/10090

Thierry J.-L. Courvoisier

From Stars to States

A Manifest for Science in Society

 Springer

Thierry J.-L. Courvoisier
Perroy
Switzerland

Translated by Stephen Lyle
Translated with the partial support of the Swiss Academy of Natural Sciences

ISSN 2191-9100   ISSN 2191-9119 (electronic)
SpringerBriefs in Astronomy
ISBN 978-3-319-59231-2   ISBN 978-3-319-59232-9 (eBook)
DOI 10.1007/978-3-319-59232-9

Library of Congress Control Number: 2017945235

Original French edition published by Georg Editeur, Genève, 2017

Printed on acid-free paper

This Springer imprint is published by Springer Nature
The registered company is Springer International Publishing AG
The registered company address is: Gewerbestrasse 11, 6330 Cham, Switzerland

For Milan, Marylou, and Marlo
Hoping that the Enlightenment will continue
to illuminate their century

Foreword

Who can look up at the night sky and fail to be filled with wonder and curiosity? Astronomy is a science that seems accessible to all, especially as described by Thierry Courvoisier, and its exploration gives the reader the rare opportunity to expand their interest beyond the narrow confines of our own planet.

The first four chapters of this book masterfully describe how science has unpicked many of the mysteries around us to reveal some of the beauty of our Universe. For both scientists and non-scientists alike, it is a real pleasure to be walked through the nature of scientific discovery and feel the excitement of the process pouring off the page. As we read further, we also learn about the pursuit of science being a fundamental part of our culture, as rich in the inspiration it can provide as music, poetry and literature. Without doubt, we know that scientific pursuit generates knowledge and understanding but we struggle to articulate the societal impact of this knowledge and how it touches every aspect of our lives. Here the author paints us a picture of the responsibility of scientists in communicating their knowledge and translating that knowledge into real impact for understanding and solving pressing global issues such as climate change and sustainable food production for an ever increasing global population.

But again, Thierry Courvoisier points out that this cannot be achieved by scientists alone and there is an increasing imperative to engage with the broadest range of stakeholders from the public to business, policy makers and politicians. These are sometimes uncomfortable bedfellows for scientists as it requires stepping out of the 'rational' world and confronting other aspects of being human such as emotion and conviction.

This engaging book is like a story of the evolution of a science, through how knowledge is generated to promoting a rational basis for political decision making. It will leave the reader slightly breathless, provoked and wanting more.

Anne Glover
University of Aberdeen

Foreword to the French edition

This book is not a vision of our planet, somehow homing in on us from Sirius, but rather a defence of science, which inspired our fellow citizens until only recently, in the hope that it will continue to do so for some time to come. His aim is to renew the dialogue between science and society which has become something less than systematic over the past few decades, so that the relationship between the scientific community and the world of politics does not just focus on investment and profit.

According to Thierry Courvoisier in his final lecture at the University of Geneva: "Astronomy has decisively extended and enhanced the sphere of human thought. Of all the natural sciences, it is perhaps the one that has made the greatest contribution to our social, cultural, and economic lives." In his book, he returns to the example of astronomy and astrophysics, from their beginnings right up to the present day, leading as they have to the extreme sophistication of modern measuring devices. He shows us beyond all possible doubt how they have indeed come to enrich our thought.

As it gradually came into being over the centuries, science was not only confronted with ignorance. It had also to answer difficult societal questions. For science is based on reason, while society is based on values. By associating "science" with "conscience", Rabelais associated human knowledge with a notion that encompasses our subjective appreciation of our acts and their moral value. Rabelais was a contemporary of Copernicus. They were both doctors at a time when the study of anatomy was not allowed to appeal to the forbidden art of dissection, and when the Copernican revolution could not be admitted because, through religious dogma, the Earth had to be at the centre of the world. The moral values of society weighed heavily on science.

Over five centuries, human knowledge has increased considerably and the values of human societies have changed, but the interface between science and society has certainly not grown any simpler. In this context of the inexorable advance of science and technology today, we come to ask: What are the limits of science? It is one that torments society, attracted by science and at the same time distrustful, even suspicious about what could be done with this newly acquired knowledge. What are its intrinsic limits? And what limits shall we ourselves impose? Indeed, should we impose limits? And why should we wish to go beyond such limits?

In his book, Thierry Courvoisier asks us to reflect upon the various aspects of the relationships between science, society, and politics. He advocates a dialogue so that societies faced with challenges they are unable to resolve can integrate scientific knowledge while taking the time required for due consideration of the consequences.

Catherine Bréchignac
Permanent Secretary of
the Academy of Sciences (Institut de France)
Paris, France

Preface

The aim of the present book is to understand the relationship between knowledge and society, then describe ways to bring the relevant science to bear on the processes of political decision-making.

For decades during the second half of the twentieth century, science was seen as a positive, indisputable, and powerful driving force for the development of Western society. But this is no longer the case. Many today feel and express a growing skepticism with regard to knowledge and reason, or even reject them altogether. The Enlightenment of the eighteenth century seems to be losing its lustre in the twenty-first. However, reason and knowledge are essential for the wise and responsible management of our planet.

We must counter the sometimes hostile approach to science displayed by some of our contemporaries. With reference to astronomy, the science I know best, the first part of the book shows just how much our society has gained, and continues to gain, culturally, economically, and politically by the development of knowledge. Science is the work of women and men. Their task alternates between routine and the search for harmony. It begins with the birth of an idea and ends with a universally recognized piece of knowledge, in principle accessible to all, and freely available to guide human endeavour.

Scientific practice involves the responsibility of research scientists to make their knowledge genuinely useful to society. The second half of the book illustrates how important it is to gain knowledge of our environment, and hence understand our own influence upon it. It then discusses the responsibilities that fall specifically upon the scientist. The women and men of science can, and indeed must, make their knowledge available to society in a form that can facilitate well informed political decisions. The dialogue between science and politics is not easy, but it has assumed a new and richer structure since the 1990s.

Most political decision-making processes take place on the national level, whereas science knows no frontiers and the planet is a single ecosystem. We must therefore find a way to go beyond this rather parochial way of thinking and establish continental and global institutions that can implement the regional and worldwide projects made essential by human influence on the planet as a whole.

Emotions and beliefs are the irrational components of our decision-making processes. Reason and knowledge must also allow us to take these fundamental features of humanity into account. Today, while human society affects the physical, chemical, and biological parameters of the planet over the whole of its surface, we must use the tools available to us to guarantee harmonious living conditions for all its many and varied inhabitants.

Tromsø, Norway   Thierry J.-L. Courvoisier
December 2016

Acknowledgements

This book was written during two stays outside Switzerland. The first was to the University of Crete in Heraklion, the southernmost university of Europe, and the second to the Arctic University of Norway in Tromsø, the northernmost university of Europe. My thanks go to my hosts in these two universities, Prof. Nikos Kylafis, Prof. Trygve Johnsen, and the rector in Tromsø, Prof. Anne Husebekk.

Many of the ideas presented here are the result of reflections while carrying out my functions in the European Astronomical Society (EAS), the *Académie suisse des sciences naturelles* (SCNAT), the *Académies suisses des sciences* (a+), and the European Academies Science Advisory Council (EASAC). I am grateful to my colleagues in these institutions for many formal and informal discussions. I am also grateful to many organisations and associations which, by inviting me to speak or write for them, gave me the opportunity to extend and enrich my thoughts on these matters.

I am grateful to Prof. J. Bell-Burnell, not only for having discovered pulsars, but also for her collection of poems relating in one way or another to astronomy. I thank her for allowing me access to this work. Thanks also to Prof. A. de Pury for a fruitful discussion on the names of God and eternity. My understanding of the role in time measurement played until recently by observatories in the Jura results from discussions with Prof. M. Golay, Prof. F. Rufener, and J.-F. Bopp in Geneva. My excursion into literature would not have been possible without comments from Prof. P. Lombardo, whom I must also warmly thank.

Professors M.C.E. Huber and D. Monard read and commented on a first version of the manuscript. I thank them for their encouraging and constructive remarks. The transformation of a simple text into a book involves a great deal of work, carried out here with great diligence and support by Michael Balavoine and Anthony Chenevard at Georg éditeur in Geneva.

Writing takes time, energy, and concentration. Thanks to my wife Barbara for her understanding and for having supported and accompanied a project carried out in sometimes rudimentary settings, in the dead of night, for example, aboard our boat Cérès in the port of Tromsø.

Contents

Chapter 1
Astrophysics Since the Middle of the Twentieth Century

The Universe that we could still discover as children in the 1950s, by reading or listening to our elders, comprised the planets of the Solar System, the Sun, stars, and galaxies. Our own star, the Sun, is just one of 100 billion others making up the Milky Way, the galaxy it belongs to. The planets gravitate around the Sun and a few moons, or natural satellites, orbit these planets, just as our own Moon revolves around the Earth. We knew the mass, the distance from the Sun, and the size of the planets and their moons, but little else apart from this handful of numbers. It was known at the time that the stars were made of gas and that nuclear reactions in their core transform hydrogen into helium and thereby release the energy that allows them to shine. The main lines of the evolution of these stars were also known. We knew that some galaxies closer than others are bound together into immense clusters, but without understanding why these systems should have remained distinct from their surroundings since the beginning of time, without dissolving into the rest of the cosmos. The expansion of the Universe had been clearly established. It seemed likely that the Universe must cool as it expands, whence it must have been much hotter in the past than it is now.

This picture had been acquired through observation by telescopes, instruments often situated on hills or mountains, in places where the sky was clearer than in our cities. Most of the modern telescopes of the day recorded the light from stars on photographic plates, which replaced the naked eye and hand-drawn sketches of the previous centuries. Much intelligence had been involved in deducing the properties of the stars from the available observations, and defeating the skepticism of positivistic thinking which, in the nineteenth century, had led Auguste Comte to say that we could never know what was going on inside the stars, for these regions would remain forever inaccessible to in situ measurement.

© Springer International Publishing AG 2017   1
T.J.-L. Courvoisier, *From Stars to States*, SpringerBriefs in Astronomy,
DOI 10.1007/978-3-319-59232-9_1

Light and Electromagnetic Waves

Ground-based telescopes and their instrumentation record visible light, the light which contains most of the energy reaching us from the Sun. Our atmosphere is transparent to it, and of course our eyes are sensitive to it because they have evolved that way: what good would they be, eyes like ours, if there were no sunlight or if the atmosphere were opaque!

Light is an electromagnetic wave. It comprises an electric field and a magnetic field, which oscillate perpendicular to one another in the plane transverse to the direction of propagation of the wave. The speed of light, like the speed of all other electromagnetic waves, is 300,000 km/s in vacuum. The frequency of these oscillations (how many there are every second) determines their wavelength (the distance between two maxima of the oscillation). The colour of light is our way of perceiving its wavelength. Red corresponds to a longer wavelength than blue. It is a happy coincidence that most of the Sun's light is emitted in colours for which the atmosphere is transparent. If this were not so, we would live in a luminiferous fog in which we could not distinguish objects one from the other. Our vision and the whole of the evolution of life would have been very different.

The oscillations of electromagnetic waves are not restricted to the frequencies corresponding to our colours. Radio waves and infrared radiation are also electromagnetic waves, just like visible light, differing only by the fact that their frequencies are much lower and their wavelengths much longer. Likewise, ultraviolet radiation, X-rays, and gamma rays are also electromagnetic waves, but with much higher frequencies, increasing as one goes from the ultraviolet to gamma rays. The frequency of electromagnetic waves determines their energy. Radio waves, with low frequencies and long wavelengths, carry little energy, while gamma rays with their very high frequencies and very short wavelengths are highly energetic (Fig. 1.1).

There is no reason why the physics taking place in the Universe should be limited to phenomena emitting only visible light. In the Cosmos, as here on Earth, visible light is emitted by objects with temperatures of the order of a thousand or a few thousand degrees. The Sun and stars emitting visible light have surface temperatures of this order, something like 6000 K for the surface of the Sun.[1] Infrared radiation is emitted by surfaces at a few hundred degrees, such as the hotplate in the kitchen, or interstellar dust. Having said this, it is easy to imagine objects in the Cosmos whose temperature is lower or higher than the surface of the Sun, for which the main part of the radiation would then lie outside the range of electromagnetic waves detectable by our eyes and our ground-based telescopes. Hence the interest in carrying out astronomical observation outside the visible part of the spectrum, something which became a real possibility in the 1960s.

Two major developments have allowed us to make observations of other kinds of light than what is perceived by our eyes and the photographic plates of telescopes.

---

[1]Throughout the book we shall use degrees on the kelvin scale, denoted by a capital K. To obtain the temperature in degrees kelvin from the temperature in degrees celsius, it suffices to add 273.

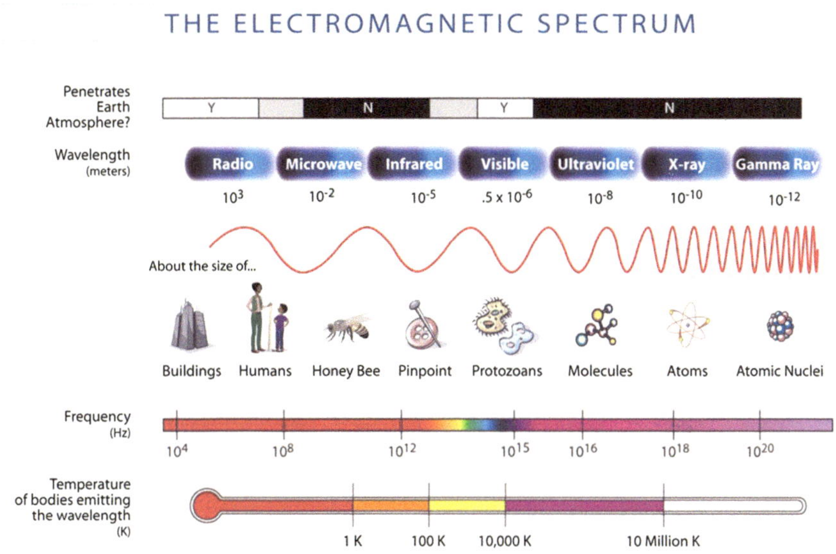

THE ELECTROMAGNETIC SPECTRUM

Fig. 1.1 Electromagnetic spectrum. Low frequencies correspond to long wavelengths and low energies. Conversely, high frequencies mean long wavelengths and high energies. *Source* Wikicommons CC. https://commons.wikimedia.org/wiki/Template:Potd/2008-04#/media/File: EM_Spectrum_Properties_edit.svg.

The first was the development of wireless telecommunications and radar, the second the possibility of sending rockets beyond the Earth's atmosphere. The first opened the way to the observation of radio waves from the Cosmos, and the second made it possible to place astronomical observation equipment in orbit around the Earth, covering the infrared, ultraviolet, X-ray, and gamma-ray regions. It was in the 1950s and 1960s, to a large extent thanks to the requirements of the military during the Second World War, that these two kinds of technology became mature enough to observe the sky. The ensuing discoveries went far beyond anything anyone expected.

New Discoveries in Astrophysics

In 1962, R. Giacconi, an Italian physicist working in the United States, led a team that launched a rocket equipped with X-ray detectors. There was no question yet of putting a satellite in orbit around the Earth. This was a flight that would last just a few minutes above the layers of the atmosphere that block out X-rays. The idea behind the mission was to observe the Moon and measure the X-rays emitted by its surface in response to the intense bombardment of particles from the Sun. This radiation was not observed, being too weak to be detected by the instruments of the day. However, a powerful source of X-ray radiation passed briefly across the field of view of the instrument. It was a source that nobody had ever imagined could

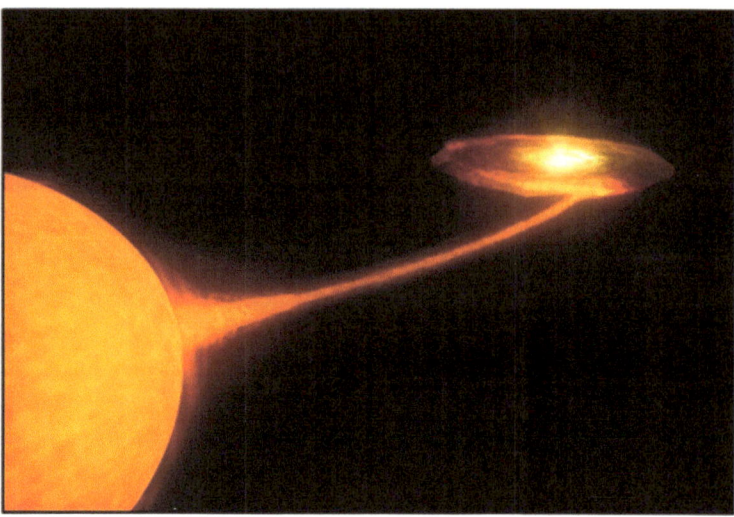

Fig. 1.2 Artist's impression of a system comprising a normal star and a neutron star dragging matter from it. The matter heats up enough to emit X-rays as it falls. Artist's impression of a pulsar 'eating' a companion star. *Source* NASA/Dana Berry ©

exist, considerably brighter than what would be expected from any star like the Sun or hotter, at the distance of the nearest star. This completely unexpected circumstance was followed over the next few decades by the discovery of thousands of other celestial sources emitting considerable amounts of X-ray radiation.

Figure 1.2 shows what is going on in this first source, and indeed in a significant fraction of X-ray sources in galaxies. These are systems of two stars, one more or less normal, the other highly compact, in fact a neutron star. The latter are composed mainly of neutrons, one of the components of the atomic nucleus. These neutron stars have masses close to the mass of the Sun, but they measure only about ten kilometres across.[2] With this mass and such a small size, the density of these stars[3] is so great that a teaspoonful of this material would contain as much mass as the whole of humanity.

The normal star and the compact one orbit around one another. When the distance between them is small enough, the gravitational field of the neutron star will become strong enough to remove some of the matter on the surface of the normal star and drag it onto the neutron star. As it falls toward the compact object, the matter heats up, just like the remains of satellites upon re-entry into the Earth atmosphere. But when matter falls onto a neutron star, the temperatures reached are high enough for the emission of X-rays, and this is what we observe. These systems can emit up to 100,000 times the energy of the Sun.

---

[2]While the Sun has a radius of about 700,000 km.

[3]About 10^{15} g/cm^3.

In some cases, the compact object is still more dense than a neutron star. Nothing can then stop it from contracting and it eventually becomes a black hole, a body so dense that not even light can escape from it.

In the same period, the end of the 1950s and the beginning of the 1960s, technological developments made it possible to measure the positions of radio wave sources (or more concisely, just radio sources) accurately enough to actually identify some of them as visible "stars" on photographic plates. Objects emitting radio waves at the same time as visible light are rare, because normal stars emit very little in the radio region of the spectrum. Moreover, some of these turned out to have remarkable properties: the visible light from these objects had characteristics that could not be understood in terms of the properties of the atoms known from laboratory measurements. It was Maarten Schmidt, a Dutch astronomer working in California, who finally solved this mystery. He showed that these objects were actually much further away than all the galaxies known at the time. The light emitted by perfectly normal atoms in these sources is thus shifted toward much longer wavelengths as it travels between the source and our telescopes, a well known phenomenon known as redshift. This discovery showed that the "stars" identified here belonged to a new category of cosmic objects whose many fascinating features would gradually be established over subsequent decades. The "stars" in question are relatively bright on photographic plates. But if they look so bright and yet lie at such enormous distances, this means they must be emitting huge amounts of energy, much more than the stars, or even whole galaxies. They became known as quasars, short for quasi-stellar objects (QSO). It was soon noticed that the luminosities of quasars vary significantly over periods of months, days, or even hours, a property unknown among stars.

It turns out that quasars are black holes with masses up to several billion times the mass of the Sun. The source of the considerable amounts of energy they emit is the same as for the X-ray binary systems, namely, matter falling onto a compact object. And the phenomenology of this falling matter is extremely rich. As it falls, it heats up enough to emit X-rays, and it forms jets which escape at speeds close to the speed of light, forming shock waves which accelerate electrons to energies where they emit both X and gamma rays at higher energies than any other observed as they interact with their environment. While it is easy enough to calculate that the amount of matter falling into the black hole is of the order of one solar mass per year for a bright quasar, it is more difficult to understand all the mechanisms underlying the appearance of such objects. Figure 1.3 shows the quasar 3C 273, the first whose distance was worked out, as seen by the Chandra space telescope, an instrument observing in the X-ray domain. The black hole itself does not emit any light. It is the falling matter around it that shines so brightly. One of the bright jets emitted by the quasar is also visible.

A few years later, it was once again radio observations that led to the discovery of pulsars. Jocelyn Bell was observing fluctuations in radio sources for her doctoral thesis when she noticed that one of them was emitting a pulse of radiation every 1.3 s (Fig. 1.4). Since no other radio or visible source had ever been observed with such a feature, it was first called LGM-1 (LGM stood for *Little Green Men*),

Fig. 1.3 The quasar 3C 273 viewed by the HST space telescope. The jet is the irregular structure bottom right. *Source* NASA-STScI ©. HST Snaps Optical Jet of Quasar 3c 273, 9 September 1993, 06:00. *Source* R.C. Thomson, IoA, Cambridge, UK;C.D. Mackay, IoA, Cambridge, UK;A. E. Wright, ATNF, Parkes, Australia / ESA ©

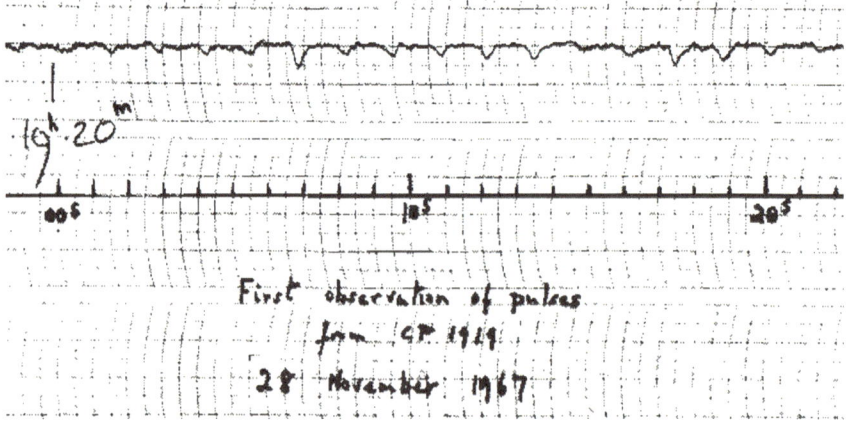

Fig. 1.4 Radio signal which led to the discovery of the pulsar, as recorded at the Mullard Radio Astronomy Observatory by Jocelyn Bell. *Source* Mullard Radio Astronomy Observatory. *Source* Jocelyn Bell Burnell and Antony Hewish ©

expressing the possibility that this signal might be the signature of an extraterrestrial civilisation. Several similar sources were then discovered and the explanation in terms of another civilisation was quickly dropped. This phenomenon is in fact due to the presence of a neutron star which spins on its own axis every 1.3 s, emitting radio waves in a narrow cone in the direction of a strong magnetic field. Since the cone is not always aligned with the axis of rotation, just as the magnetic axis of the Earth is not aligned with geographic north, we perceive a flash each time the cone sweeps across the Solar System, rather like a lighthouse beam as its optical system turns round. It is this feature that gives the name "pulsar". Before these observations, nobody would have imagined that neutron stars could manifest themselves in this way.

Thousands of pulsars have been found since 1967. Some spin with periods as short as a few milliseconds. This means that, on such an object, with mass close to the mass of the Sun, a day lasts only a few milliseconds. While these standard pulsars are above all radio sources, others onto which matter is falling emit X and gamma rays. The latter are still under investigation, among other things to get a more precise understanding of their internal composition. Indeed, the matter there is so dense that we know of no other environment, not even inside an atomic nucleus, in which such conditions prevail and might be studied.

In the second half of the 1960s, the American VEGA satellites were launched to seek out Soviet nuclear test explosions in the atmosphere. They didn't in fact detect any such explosions. However, they were sensitive, not only to gamma rays coming from the Earth's surface, but also to sources of gamma rays in space. Against all expectations, these detectors picked up sporadic bursts of gamma rays from space, each lasting of the order of a second. These gamma bursts, as they became known, occur something like once a day and come from all directions in space. They can be as short as a few milliseconds or last for a few seconds (Fig. 1.5). There is no way of predicting what direction they will come from, and for a long time only their gamma radiation was known. There was much speculation about the origin of this phenomenon, until one day at the end of the 1990s, an X-ray signal was also picked up from the same source as a gamma burst. This allowed a precise localisation of the burst on the celestial sphere, and from this it was deduced that gamma bursts are due to phenomena occurring at the same kind of distances as quasars, that is, way out at the edges of the observable Universe. Exceptionally bright and very remote, they are, like quasars, very high energy sources. Some of these are undoubtedly related to the collapse of a normal star at the end of its life, while others are most probably due to the coalescence of two neutron stars, giving rise to a black hole.

More energy is emitted in a gamma burst in a few seconds or a fraction of a second than the Sun will emit during its whole lifetime of around ten billion years. The emission is not distributed spherically around the source, but occurs rather in the form of a jet, as in the case of quasars and pulsars.

At the end of the 1970s, a different kind of gamma burst was discovered: repeating bursts which could not be due to the collapse of a star or the coalescence of neutron stars, since such an event can only happen once in any given system. This time the origin of the phenomenon is a completely different story. It is due to

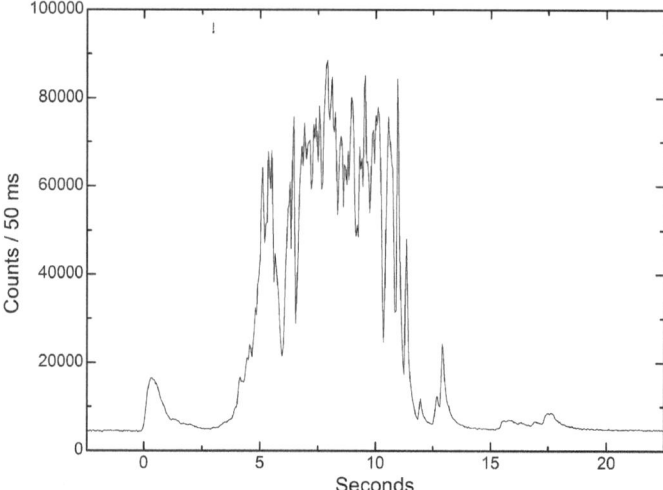

Fig. 1.5 Intensity of gamma rays recorded by the satellite INTEGRAL during a gamma burst which occurred on 27 April 2013. The burst lasted about 10 s. Such bursts are observed at a rate of about once every day

rearrangements of the most intense magnetic fields so far observed in the Universe, at the surface of a neutron star.

These broad categories of cosmic phenomena and objects each divide into a multitude of subclasses with a wide range of different features. None of them were predicted before they were actually discovered, even though the existence of neutron stars and black holes was known from a theoretical standpoint.

Once such a new category of phenomena is detected, the first task of the astrophysicist is to organise the observations, just as was originally done in zoology and botany, looking for features common to different objects, defining categories and arranging the phenomena into consistent classes. The next task is to bring to bear all available knowledge of physics to understand these phenomena and find a plausible explanation. This is a difficult undertaking for the astrophysicist because, in contrast to research carried out in the laboratory, he can only use observations of the phenomena that nature has chosen to show us. There is no way of influencing the various parameters of the system in order to identify their individual roles.

All these objects have been discovered since the 1960s. Given the variety and strangeness of the observations, combined with the fact that there are rarely opportunities to reproduce the observed conditions in the lab, all this has happened in a very short lapse of time. Few instruments have been dedicated to this work, beyond a couple of dozen UV, X-ray, and gamma-ray observation satellites. Much more observation, intelligence, and effort will be needed to get a satisfactory understanding of all the phenomena so far discovered.

Fig. 1.6 The starry sky above the European Southern Observatory (ESO) on Mount Paranal (Chile). Our own galaxy, the Milky Way, is visible in the form of an irregular diagonal. The bright trace is the track of a shooting star, one of the very few fast phenomena that can be seen on the night sky with the naked eye. The 2010 Perseids over the VLT, 16 August 2010, 10:10. *Source* ESO/S. Guisard ©

Temporal Variation Is the Rule for the New Classes of Object

The contrast between recent discoveries and other phenomena we have known about for a long time is quite striking. On the one hand the stars and galaxies evolve over time scales of millions or even billions of years, while on the other, the new types of object, involving phenomena whose energies can dominate whole galaxies, vary over extremely periods of time, even compared with the human time scale. This is particularly clear when we obtain X-ray and gamma-ray images of the sky at regular intervals, for example, over one or two days. The X-ray images of the observed region are all different, while the starry sky looks just the same from one night to the next, the only visible changes being the movements of the planets and the slow roation of the Earth around the Sun (Fig. 1.6).

Change, a Universal Constant

The apparent serenity of the starry sky has had a profound influence on our way of thinking. We naturally associate eternity and immutability with the notion of perfection. In the sixteenth century, when making the first French translation of the bible, Olivétan thus used "l'Éternel" as the name of God. It also explains why there are so many promises of eternal life in catechism. The sphere and immutability were ideals of perfection throughout Aristotelian philosophy, in which only the sublunar world was "corrupted" by temporal change and non-spherical shapes. Galileo put paid to this ideal after examining the surface of the Moon with his telescope, when he asserted that its surface was covered with mountains, like the Earth's. Our new understanding of the Cosmos acquired with modern instruments goes even further in breaking the association between ideality and immutability by showing that the whole Cosmos is in fact the scene of violent transformation. So it turns out that our own immediate environment, the Earth, far from being an exception in the Universe, is in fact thoroughly representative.

The idea that the Universe could never undergo transformation of any kind was so deeply rooted in the collective imagination at the beginning of the twentieth century that, when he noted that the equations of general relativity he had developed to describe gravity delivered no static solutions for the large scale structure of the Universe, Einstein modified these equations by adding a term proportional to the so-called cosmological constant. The discovery by Lemaître a few years later that the Universe was expanding apparently made this term redundant (at least that was the way it seemed), since it showed that the Universe is not in fact static and immutable.

The story about the inclusion of the cosmological constant to describe the Universe because it was supposed to be static, at the same time as debate was raging over the theory of evolution of the species during the nineteenth century, shows that the Aristotelian distinction between the corrupted sublunar world and the ideal world that was supposed to exist beyond the lunar sphere long remained a feature of the collective conscience. And yet the discoveries of modern astrophysics show that such a distinction cannot be upheld. The objects and phenomena discovered over the past few decades are as many demonstrations that change, often violent, is present on all scales now accessible to us. There is no privileged place in the Universe where anything qualifies for immutability.

Our understanding of the Cosmos and the phenomena we observe in it occupies a fundamental place in our culture. It allows us to put ourselves into a broader context and tell our story as part of the Universe as a whole. This story is not the same if we consider the Earth as the centre of an unchanging world or if, as astronomers have shown, it is just one of many planets revolving around just one star among billions of others in a context of constant evolution.

Modern astrophysics has shown that everything in the Universe is subject to change and that complex geometric forms play an important role. It has contributed to a profound change in the way we view the world by showing that the essence of existence lies rather in perpetual change than in any kind of immutability that has been arbitrarily declared perfect.

Chapter 2
Astronomers and Time

Astronomers of the past few decades have above all been the authors of discoveries that have revolutionised our vision of the Universe. They have developed instruments and used them to observe the sky, measuring quantities that were accessible to them such as the temperature of a medium, the energy of particles, or the chemical composition of a gas. They have developed models which provide a consistent picture of these measurements, then used their models to predict new properties of the given objects, and so on and so forth. The whole of modern astrophysics is characterised by this way of doing things, which already goes back to ancient times. It is the essence of the scientific method, applied to observation of the sky.

Until the beginning of the seventeenth century, all instruments used the naked eye as the only optical and detecting system, in the first place to measure angles. What was studied was the place of the Earth in the Solar System and in the Cosmos, and models of this Cosmos, called cosmogonies. These observations took a great step forward with the construction of Tycho Brahe's great observatory on the island of Hven in the strait between Denmark and Sweden, or the Jaipur observatory in India. Then Galileo turned his refracting telescope to view the Moon, the Sun, and Jupiter, opening the way to a closer analysis of the nature of celestial bodies. However, this kind of research would long remain a minor role for astronomers. Until recently, their main activity was to measure time, on the scale of seasons and years, and then on ever shorter scales of days, hours, minutes, and finally seconds and fractions thereof.

Time, Apparent Motion of the Heavenly Bodies, and Mechanics

Going back as far as records can go, humans have used the apparent motions of the heavenly bodies to measure time go by. The Sun, the Moon, and the stars provide the only points of reference for gauging the passage of time without having at least

© Springer International Publishing AG 2017
T.J.-L. Courvoisier, *From Stars to States*, SpringerBriefs in Astronomy,
DOI 10.1007/978-3-319-59232-9_2

some elementary knowledge of the laws of physics. Celestial objects that can be perceived by their visible emissions are stable enough to be used to this end. In a world where observation of the sky was done in the X-ray region of the spectrum, where objects constantly change in intensity and sometimes even disappear, it would probably be impossible to use the observed changes to take stock of passing time.

Other regularities than the celestial motions, such as the regular swing of a pendulum or the vibration of a crystal, or indeed atomic transitions, cannot be exploited to measure time without a solid grounding in the phenomena which govern such systems, an understanding that we have only acquired relatively recently.[1] The oscillations of pendulums were not understood by savants until around 1600, and it was only then that their motions could be used to measure time in clocks, indeed, in pendulum clocks. Later, crystal vibrations replaced mechanical ones in most modern watches. And in recent decades, it has been possible to control oscillations associated with atomic transitions well enough to use them as the basis for time measurement in our own society.

The Seasons

There would be no way to follow the cycle of the seasons without recognising the apparent motion of the Sun during the year. Every day at noon it gets higher in the sky as winter and spring go by, but then comes back down little by little during the summer and autumn, at least in the northern hemisphere (but the opposite in the southern hemisphere). The first monuments erected by humans and which have remained to this day, standing stones set in place long before the advent of writing, like the Standing Stones of Stenness (Fig. 2.1) in the Orkneys or Stonehenge in England, were sometimes aligned with certain important moments of the astronomical year, such as the winter solstice at the time of their construction. So it was skywatchers, astronomers of a sort, who presided over the construction of these monuments, which could then be used to determine the seasonal cycles throughout the year.

Determining the direction of the winter solstice in the British Isles during prehistoric times would certainly have taken years of painstaking observations, memorising positions and reporting the results from one year to the next. Once the geographical direction of the solstice had been determined, it would probably still have been necessary to convince the community that it was worth setting up monuments along this axis, as this construction would have represented a considerable work load for such societies. Today we would say that it represented a significant fraction of the community's gross domestic product. If we judge by the difficulty we have in convincing our own authorities of the need for our work, the

---

[1]We have made no mention of the hourglass and the clepsydra.

Fig. 2.1 Standing Stones of Stenness on the main island of the Orkneys. *Source* S. Kessler ©
https://drinkvicariously.files.wordpress.com/2013/08/dsc05015.jpg

task of persuasion required by our ancestors to obtain such a result must have been
forbidding!

Once erected and aligned, these monuments could be used to measure time
during the year. This was a crucial source of information for the practice of agriculture. Indeed, in order to plant seeds in the right season, it was essential to follow
time on a monthly scale. Meteorological time is not a reliable indicator for this, as
we all know. It was observation of the sky, the Sun, or the stars, the work of
astronomers, which allowed prehistoric societies to develop agriculture. The ability
to measure time during the year was essential for the development of civilisation,
and it is one of astronomy's contributions to human development.

Time and Organisation of Society

Determining the seasonal cycle is a central problem for agriculture, but it is still not
accurate enough to arrange a meeting somewhere, a key aspect of social organisation. To do that, we need to be able to measure the passage of time during the day.
Even there, it is down to astronomy to take up the main part of the challenge.
Diurnal time is determined by following the Earth's rotation about its own axis.

And this is measured from the Earth's surface by observing the transit of stars at the meridian, that is, the plane specified by the axis of rotation of the Earth and the position of the observer. The time required for one rotation of the Earth about its axis, the day, is given by the time between two consecutive transits of the same star at the same place. The Sun itself crosses the meridian every day. This is the definition of noon. But in one day, the Earth has moved a certain distance on its orbit around the Sun, in fact, something like one degree (about 1/365 of the Earth's yearly journey around the Sun). Viewed from the Earth, the Sun has thus moved through the same angle relative to the stars. The lapse of time between two consecutive noons, called the solar day, is thus not the same as the lapse of time between two transits of the meridian by a distant star, the so-called astronomical day. These details had to be gradually understood and taken into account to be able to provide accurate timekeeping in a given community and thus facilitate its organisation.

Astronomers invented the instruments needed for their observations using the technology of the day. They then carried out the necessary observations night after night for centuries, in all those communities that considered a mastery of time important for their internal organisation. The effort involved in this work represents one of astronomy's central contributions to modern life, even if this particular contribution has become much less important in recent decades. To judge by the mandates given by the authorities to their observatories over the last few centuries, in many towns of varying size, time measurement came high on the list of an astronomer's duties. The latest advances in astrophysical understanding were a secondary activity in these observatories, at least until the middle of the twentieth century.

Just communicating time to a given population is another problem that needs to be solved. While the local astronomer may know, for example, that it is noon at his observatory, he can hardly set off on foot or on horseback to tell the mayor or some other authority that it is noon. It would no longer be noon by the time he arrived. Since sound travels at about 300 m/s, sound signals were not accurate enough to carry information about time, for example, to meet the needs of ocean navigation. The city of Athens, and others too, got round this problem in the nineteenth century by using a visual signal. The observatory determined noon and lowered a ball down a mast placed on the roof of the observatory at exactly this time (Fig. 2.2). The mast was visible from the cathedral, so the bellringer could accurately sound the bells and hence inform the whole population of the time every day. The delays due to the propagation of sound are short enough not to pose any problem for the purposes of everyday life, since nobody needs to meet their appointments with an accuracy of seconds. But the same is not true for navigation.

For several centuries and up to relatively recently, observatories were predominantly there to serve their communities through the provision of high quality time measurement. For example, the observatory in Besançon (France) was founded in the 1870s, but it was not set up so that astronomers could contribute to an understanding of the Cosmos. The aim was rather to support the local clockmaking industry which found itself at a certain disadvantage with respect to its competitors

Fig. 2.2 The Athens observatory, founded in 1842 to provide time to the city. Every day at noon, a ball was lowered on the mast. *Source* Fingalo, Published under the Creative Commons Attribution-Share Alike 2.0 Germany license. https://commons.wikimedia.org/wiki/File:07Athen_Observatorium1.jpg

in Geneva and Neuchâtel, for the latter had their own chronometric observatories. In Geneva, the observatory carried out precise measurements on chronometers made in the region up until the 1960s.

From the middle of the last century, technology exploiting vibrations of quartz crystals, then another making use of atomic transitions, gradually replaced mechanical clocks and watches, providing more accurate and more economical timekeepers than astronomical observations on timescales of the order of days. But atomic time nevertheless remains linked to astronomical time, since the gradual slowing of the Earth's rotation due to friction induced by tidal effects is regularly compensated by adding intercalary seconds to civil time so that it remains in phase with astronomical time.

Time and Navigation

The measurement of time and the measurement of position on the Earth's surface are closely related. Time is given by the transits of celestial bodies across the meridian, and the meridian is given by the position of the observer. The difference in longitude between two points on the Earth is determined by measuring the time difference between the passage of a star across the meridian at each of the points. Moreover, the height of the Sun above the horizon at midday, a measurement used to establish the observer's latitude, depends not only on the position of the observer, but also on the day when the measurement is made. So knowledge of the relative position of two points on the Earth depends, like the measurement of time, on astronomical measurements.

Such knowledge underlies all geographic cartography, and it also underlies all navigation when there are no recognised visual points of reference, and in particular, when crossing the oceans. The navigator measures progress by regularly determining the ship's position relative to the Sun and stars and referring to a pre-established map.

It is only since the 1990s that we have been able to determine our exact location by receiving signals generated by satellites in orbit around the Earth, satellites belonging to the American GPS system and the European Galileo system (still under development), or other satellite systems. The knowledge we have of a given location through a global positioning system depends in turn on the knowledge we have of the orbits of the satellites and their progress along these orbits. These quantities can be measured geometrically from points whose positions on Earth are precisely known. However, the latter positions must be determined independently, and this is done by making astronomical measurements. Consequently, as for the measurement of time, the most up-to-date positioning technology is firmly grounded on astronomical basics.

Progress in cartography and navigation has been based on the work of astronomers ever since they have been able to make accurate enough measurements of the positions of celestial bodies to determine positions more precisely than by simply keeping records of one's journey. Knowledge of geography and navigation is also a central issue in managing and defending a territory, or indeed in taking over a "new" territory. It is essential for governments and the military, who have financed, sometimes begrudgingly, the work needed to set these systems up, and have often placed the results under the seal of secrecy.

Time and Ritual

The time measured by motions of celestial bodies is also the time used to schedule religious ceremonies. Ramadan lasts one lunar month and ends when the Moon is once again visible after a New Moon. Easter is celebrated on the first Sunday following the first Full Moon after the spring equinox. The algorithm determining this date may seem a little imprecise, and this has indeed led to some controversy, and differences between the dates of Easter in the Western and Orthodox Churches. Once again, it is astronomers who set the schedule for the religious life of a community.

Regular celestial events provide points of reference on the calendar to plan festivals and other celebrations, while unexpected celestial events, such as the appearance of comets, have also played a role in communities who endow them with premonitory meaning, often treating them as bad omens. As a matter of fact, cometary trajectories are just as predictable as planetary orbits, but since they spend most of their time a very long way from the Sun, invisible to us at such distances even with our most impressive telescopes, comets give the impression of coming mysteriously from nowhere. Their appearances were long interpreted as signs announcing important events. Likewise for eclipses, alignments causing the Sun to disappear momentarily behind the Moon. These were considered as omens of various kinds until it became possible to predict them with perfect accuracy using the tools of celestial mechanics.

The close relationship between astronomy and time measurement is not only found in the Judeo-Christian, then Western society. The regularity of daily and annual apparent motions in the sky is a universal observation, like all scientific measurements. For example, the Aborigines in Australia named their seasons, of which there were six, after stars which appeared in the sky at the beginning of each season.[2]

Chinese astronomers made a close link between the apparent motions of celestial objects, the calendar, and political power. Since the number of days in the year isn't a whole number, and likewise the number of lunar months in the year, calendars must be regularly adjusted. We do this by adding a day every four years, at the end of February, thus defining what we call a leap year. A smaller correction must still

---

[2]Norris R.P. and Hamacher D.W., Proceedings IAU symposium 260, D. Valls-Gabaud and A. Boksenberg, eds, p. 40.

be made by omitting one leap year every hundred years. Chinese calendars lost their original relationship with the seasons over the decades and had to be readjusted from time to time. Such adjustments were carried out at moments of important political change. A new beginning in the calendar would thus be associated with a new administrative period.[3]

Astronomy in the Service of Society

The ability to measure time and determine one's location were of great importance for agriculture, the organisation of society, and navigation. State authority of whatever form and wherever it might be thus depended on astronomers carrying out several of their key missions. This privileged role of astronomers in state and societal organisation still gives astronomers a position of eminence in the hierarchy of the United Kingdom, under the title of Astronomer Royal. The fact that the astronomical knowledge we have acquired over recent decades is more closely related to the physics of cosmic objects than to the measurement of time for societal purposes means that this position, or its counterpart in other nations, is now merely a matter of prestige, no longer actually endowed with real power as it used to be.

Astronomers have thus provided a very real service for the public good and the governments they have worked for over the centuries and millennia. This service was a key resource for the functioning and development of collectivities up until the second half of the twentieth century. The work done in this respect very largely dominated all the other activities of most other astronomers. The acquisition of knowledge which has since become the central activity of the astronomical community is also a form of service to humanity, although now more abstract. Human culture has grown richer throughout the history of humanity, and continues to do so, through knowledge acquired by our understanding of the sky and the objects we discover there.

---

[3]X. Sun, Proceedings IAU symposium 260, D. Valls-Gabaud and A. Boksenberg, eds, p. 98.

Chapter 3
Astronomy and Modern Society

Astronomy helped to build civilisations through its intellectual contribution. And it is still doing so. It has given humans the tools needed to measure time and space on Earth. For decades now, it has also impacted the modern world through its profound technological and political influence.

Astronomy and Technology

The first astronomical discoveries of the 1960s, including quasars, X-ray binary systems, and pulsars, came by applying technologies developed in a quite different context, namely the military requirements of the Second World War. To a large extent the technology used to develop radiotelescopes in the 1950s resulted from the construction of radar systems by the Allies to detect aircraft. The rockets used to carry the first X-ray detectors beyond the Earth atmosphere were derived directly from the German V2 rockets used to bomb the towns and cities of England.

The wealth of discoveries made possible by these transfers of technology from military to civilian science, and the will to understand the phenomena thereby revealed, subsequently replaced the previous purely military motives, leading to the development of even more advanced observation techniques. Scientists involved in the new areas of astronomy, which would eventually become high energy astrophysics, soon began to devise successive waves of ever more sensitive instruments. X-ray astronomy provides a good illustration of the way things happened. Having discovered a first bright X-ray source, it was natural to ask whether there were others, and to try to identify their properties. What was needed were instruments that were more sensitive to the X-ray flux than the first generation, and able to observe the whole sky. But while it is important to be able to measure the X-ray flux, this is still not enough to understand the physical phenomena behind such emission. For example, to know the temperature of the source, the signal must be measured at several different energies. To find out what kinds of atoms are involved

© Springer International Publishing AG 2017
T.J.-L. Courvoisier, *From Stars to States*, SpringerBriefs in Astronomy,
DOI 10.1007/978-3-319-59232-9_3

in the underlying processes, the energy of the X-rays must be very accurately measured. The series of questions that usually follow the acquisition of new measurements thus lead steadily toward the development of a new generation of detectors. And so it goes on, with ever more sophisticated, and hence more expensive instruments. The largest instruments developed in the 2000s cost as much as a billion euros, dollars, or whatever (the currency is hardly relevant at these orders of magnitude).

This evolution is regularly met with questions about the meaning and relevance of such research for society.[1] However, quite apart from reflections on the opportunity to pursue research using very large instruments, astronomy has become a powerful driving force for technological development.

X-ray astronomy, focusing on observation of the X-ray region of the electromagnetic spectrum, has come a long way through a series of space missions starting with the first flight in 1962 and building up to the telescopes launched in the 2000s. Figures 3.1 and 3.2 give some idea of the size of these missions today as compared with those of the 1960s. The first such missions carried detectors with a mass of a few kilos above the atmosphere for a few minutes, while the most recent place telescopes with masses of several tonnes in orbit outside the atmosphere for ten years or more. The size of these instruments and telescopes, and their long lifespans, are certainly factors that increase the complexity of such missions, but they are far from being the only ones. Detectors are reaching the limits imposed by fundamental physics. The problem is to measure, for each photon, each grain of light, the direction it comes from with an accuracy determined by the principles of optics, its energy with a precision dictated by atomic physics, and the moment of arrival of the photon in the detector with the accuracy of an atomic clock.

In the optical region of the electromagnetic spectrum, photographic plates were used as detectors up until the 1980s. The largest telescopes had a diameter of three to five metres. Their mass and hence their diameter was limited by the constraint of mechanical stability. Today detectors are electronic, and often astoundingly complex. Electronics is used to control the shape of a mirror in an active way, and we can now go beyond the limits imposed by the need for the telescope to be structurally rigid. Today's telescopes have diameters of 8–10 m, and a telescope of diameter 39 m is currently under construction at the European Southern Observatory (ESO), set up by Europe to take full advantage of one of the most favourable sites in Chile (Fig. 3.3).

As astronomers make more and more discoveries and subsequently seek to further their understanding, their requirements are becoming increasingly demanding, and this has become a powerful driving force for technological innovation. The aim is to open up new areas of astronomical observation, and to obtain in every case a maximum of information with a minimum of observation. To do this, one must come

---

[1]See *Space Sciences, Why? Three essays to explore the value of science in Space: Earth, Exploration, Stars,* J. Remedios, E. Messerschmid, S. Wittig, T.J.-L. Courvoisier, 2016 (http://fr. calameo.com/books/0011383688fc19f0a0abe).

Fig. 3.1 An Aerobee rocket similar to the one used by R. Giacconi and his colleagues for the flight in which the first extrasolar X-ray source was discovered. *Credit* USAF

as close as possible to the fundamental physical measurement limits, rather than tolerate technical imperfections, such as defects in the quality of optical surfaces or electronic components. Research programmes are extremely ambitious and the practical demands considerable. Astronomical observation is always carried out under difficult conditions, in the great outdoors or in the emptiness of space. Instruments must be designed to work in hostile environments like the Chilean desert, the Antarctic, or outer space. Spaceborne instruments must resist the stresses induced during launch, and they must be exceptionally reliable so that they can operate for years without human intervention.

The complexity of an instrument can be assessed through its cost. Large modern instruments and scientific satellites can reach hundreds of millions of euros, or even a billion. A cost of a billion euros corresponds to 10,000 years' work, if we estimate

Fig. 3.2 Launch of the X-ray observation satellite XMM-Newton by the ESA in 1999. Ariane 504 's launch, 10.12.1999 10:51 am. *Source* ESA/CNES/Arianespace-Service Optique CSG ©

Fig. 3.3 The MUSE detector used on one of the 8 m telescopes at the European Southern Observatory (ESO) in Chile. The MUSE instrument on the VLT: equipped with 24 continuous flow cooling systems, 9 June 2015, 17:00. *Source* ESO ©

that one year's work costs 100,000 euros. Since costs are largely dominated by the cost of manpower, this means that we are talking about the combined efforts of 1000 people working for a decade to obtain a major modern astronomical measurement instrument. This material, technical, and scientific challenge is accompanied by a considerable human and organisational challenge. Obtaining funding, finding people with suitable skills, getting teams of very different origins to work together, monitoring progress, keeping an eye on the project as a whole, and controlling all-important details are so many battles that must be won if the final instrument is to gather the required data. And yet there have been many successes and few failures. Despite the difficulties involved, teams acquire the skills needed to meet the scientific requirements.

Learning to build and operate instruments on this level of complexity and quality has been, and still is, an enriching experience for the space agencies, research institutes, and industries engaged in such projects. This considerably improves their capabilities on the scientific, technical, and even human level. The know-how of the various stakeholders involved directly in such scientific developments then overflows to all the other activities of other parts of the relevant institutions and industries, and not only those directly involved in building these highly complex instruments. Through this internal transfer of skills and technology, they can improve the technical processes and human organisation going on within. Products generated in connection with or in the wake of ambitious scientific projects are then often more successful than those of rival companies, who must therefore also acquire similar skills in order to remain competitive. In this way, the know-how derived from the tough demands of astronomy gradually diffuses into a whole range of different areas of industry. As time goes by, the actors in such projects move through the institutional and industrial fabric, contributing in another way to the spread of expertise acquired in the context of major scientific projects. Society as a whole thus benefits from know-how obtained by pursuing state-of-the-art scientific objectives.

In some areas, the contributions of astronomy to society are quite tangible. An example is medical imaging using X-rays. These images are produced using the differences in absorption of X-rays by different body tissues. Bone absorbs more than muscle and this appears quite clearly in radiological images. But X-rays are dangerous so it is essential to minimise the patient's exposure as far as possible while still obtaining high quality images. As in the field of astronomy, the problem is thus to extract a maximum of information, in this case an image showing the part of the body crossed by the X-rays, with a minimum number of photons. Techniques for detecting and analysing data developed for the construction and exploitation of instruments carried aboard an X-ray observation satellite here provide an invaluable model.

Practical Uses for Space Technology

Space technology is not only used to observe the Cosmos, but also to monitor events in and around the Earth itself, and to develop tools that have now become everyday items. Weather forecasting would be unthinkable today without imaging or regular measurements obtained from space. Satellite positioning systems have become indispensable for navigation, by road, by sea, or by air. Spaceborne telecommunications are ubiquitous. Monitoring and management of our urban, natural, and military environment have become commonplace. In our daily activities, each of us can find a role of some kind played by the use of satellites. And what is true for each of us is also true for the economy in general. Science has been a crucial driving force in these developments.

Since the 2000s, however, space agencies have ceased to consider science as a motivation for the development of space technology. They build their strategies around what are essentially utilitarian and commercial factors. A review report on the scientific programme of the European Space Agency (ESA) in 2007 employs a past tense: "Space science in Europe has initially been the main driver for the development of space technologies, which were later the basis for many applications serving a wide range of societal needs".[2] Science is thus no longer viewed as the main driving force in the development of space technology, only as one client among others for services and tools provided by the major space companies.

---

[2]ESA/C (2007) 13, January 2007.

This assessment puts a stranglehold on the funding of modern space science and weakens the dynamics generated up to now between the scientific world and the world of services and industry. Scientific incentives are growing less common. They no longer motivate, or only marginally, the development of technology that can benefit all of us.

We have acquired a whole range of spaceborne tools that have transformed our way of life and without which our standard of living would be significantly reduced. This dependence of our society on space techniques, like any dependence, makes us vulnerable. This means that we should pay particular attention to the maintenance and development of these tools, and the know-how needed to maintain and develop them. Considering space technology purely as a commercial asset rather than as an integral part of our modern civilisation may have very serious consequences. A striking example here is the time it took Europe to set up its own satellite navigation system. While the GPS was developed by the US in the 1970s and 1980s for military purposes before being made available for civilian applications, Europe's Galileo system is still not available to the general public in 2016. This means that Europeans are dependent not only on foreign space technology, but also on the willingness of the North American authorities to allow them access to these facilities. A vigorous scientific space programme would allow Europe to rebuild a constructive dynamic between the novel demands of science and the industrial capabilities needed to satisfy them. And it would thus in part make up for lost time with respect to other space programmes around the world. Europe would thereby reduce its dependence on other world powers by pursuing its own ambitious scientific space policy.

Science Is One Technological Driving Force Among Others

It would be naive to think that scientific requirements are the only factor contributing to progress in space technology, or even absolutely necessary for astronomical observation. Galileo did not design the refracting telescope. He merely turned an instrument designed for terrestrial, and in particular military use, in the direction of the sky. Infrared light detectors, sensitive to electromagnetic radiation emitted by objects at temperatures of a few hundred degrees kelvin (0 °C corresponds to 273 K), hence by the human body, were very much of military inspiration. Mass-produced electronics is another powerful driving force for the development of light sensors and for a whole range of products and software that bring great advantage to science. The processing, analysis, and archiving of data gathered by the astronomical observation satellite INTEGRAL (in the gamma region) provides a good example: the satellite generates something like a terabyte of data every year. This was a "big data" project at the beginning of the 1990s, when it was designed. But data storage and transfer technologies evolved so much during the 1990s that the data from this satellite can now be processed by a few computers and disks, considerably simplifying the technical and scientific work needed for the project. It is not the requirements of the research scientist that have determined and will determine this evolution, but rather the amounts of data that most of us now produce when we use our mobile phones and other devices to generate and share photos and videos.

The technological evolution of our society results from the requirements of society as a whole, the scientific community, and military systems. These three actors continually express more or less openly and explicitly the desire for applications that go beyond what is already acquired or possible at any given time. The effort put into satisfying these demands serves not only those working in the area they are originally intended for, but subsequently benefits everyone, and hopefully improves the lives of all concerned.

Astronomy and Geopolitics

The connections between astronomy, science, and modern science are also political. It is striking to read the speech made by J.F. Kennedy in Houston in 1962 to launch the race into space, and in particular to the Moon.[3] There are of course scientific ambitions: "We have vowed that we shall not see space filled with weapons of mass destruction, but with instruments of knowledge and understanding." But these are mixed with aspirations of foreign policy: "For the eyes of the world now look into space, to the moon and to the planets beyond, and we have vowed that we shall not see it governed by a hostile flag of conquest, but by a banner of freedom and peace." And the desire for leadership and domination: "Yet the vows of this Nation can only be fulfilled if we in this Nation are first, and, therefore, we intend to be first." As Kennedy says, not forgetting a reference to industrial competitiveness, all these things are closely related: "In short, our leadership in science and in industry, our hopes for peace and security, our obligations to ourselves as well as others, all require us to make this effort, to solve these mysteries, to solve them for the good of all men, and to become the world's leading space-faring nation".

The strange mixture of scientific, industrial, military, and political strategy expressed in this speech was in no way diminished when the first men, Americans, set foot on the Moon. The Chinese space programme, which has gained in scientific and technological importance since the beginning of the 2000s is another example. As with the Americans and Russians previously, this programme has as many political and industrial connotations as scientific ones.

The major European astronomical and space programmes have been much less often staged as a demonstration of power than their counterparts around the world. This relative modesty in the image they choose to present contrasts starkly with the technical and scientific ambitions that motivate them. For example, Europe leads

[3]Text of President John F. Kennedy's Rice Stadium Moon speech, 12 September 1962 (https://er. jsc.nasa.gov/seh/ricetalk.htm).

the world in astronomical research with the large telescopes of the European Southern Observatory (ESO). However, Europe is not a national power and has no ambitions of domination. Indeed, it does not cultivate any form of nation worship or patriotism, the prerogative of individual nations. It is individual countries that each try to dominate others, not the supernational entities which these states have no intention of allowing to shine in their place. The tensions between national endeavour and efforts made by the continent as a whole, combined with the lack of European geostrategic arguments, are largely responsible for the modest position of the European space programme on the world stage. With no sentiment of national prestige or military space programme to drive it forward, Europe is condemned to remain a second rate space power, guaranteeing its continued dependence on the major powers for a significant part of its standard of living. And this despite an original scientific programme maintained by highly competent people and organisations as good as any anywhere in the world, and despite the fact that Europe is the world's leading economic power.

Ground-based European astronomy suffers less from the rather unconvincing image of its spaceborne programme because it has been under constant development at the European Southern Observatory (ESO) since the 1960s. It started out from a very modest position inherited from the Second World War and succeeded in setting up four 8 m telescopes equipped with state-of-the-art instruments on Mount Paranal in Chile. Needless to say, such a development did not come without disagreements between member countries, but the relative lack of importance attributed to astronomy as an element of European geostrategy certainly helped it to avoid becoming the victim of sterile nationalistic ambitions.

Removing geostrategy from the relationship between science and society leaves the possibility of synergies with industry and innovation. This relationship is not restricted to the transfer of technology and know-how stimulated by scientists' new and ever more accurate measurement requirements. It also involves the training provided by scientists in the universities, the contributions of research scientists who go into industry after years of academic research, and scientists' efforts to share their knowledge with the economic world. It is hard to quantify what science brings to society in economic terms, but one study[4] suggests that this contribution may very largely dominate technological transfer in the strict sense of the term.

Astronomy is a magnificent branch of science. Its interactions with society are very rich, partly because of the fascination many of us have for the night sky. However, it remains just one of many scientific disciplines, all of which bring culture, knowledge, technology, and geopolitical ambitions in various proportions and with a different historical background, and each of which also has its heralds, more or less well known to the general public. Biology has made fundamental contributions to human culture, for example, by showing that all the species,

---

[4]Cornelis A. van Bochove, *Basic Research and Prosperity: Sampling and Selection of Technological Possibilities and of Scientific Hypotheses as an Alternative Engine of Endogenous Growth*, CWTS-wp-2012-003 (http://hdl.handle.net/1887/18636).

including humans, have evolved over the ages. It has brought with it an impressive technology which modifies and prolongs our lives, and it plays an important geopolitical and military role. Likewise for physics, which has shown that neither space nor time are absolute, and has revealed the mysteries of quantum theory. Physics has contributed even more profoundly to our everyday electronic environment. We only have to think of the telecommunications devices we use on a daily basis. It also remains an important tool for military developments. Chemistry and geology are equally familiar to us in this context, as much for the discoveries that have transformed our understanding of the constituents of matter (the periodic table of the elements) and the Earth, as through practical applications and contributions to political debate, for example, when we appeal to geology in discussions over the "ownership" of natural resources on the ocean floor. The human and social sciences, although less quantitative, nevertheless contribute to our understanding of the world, and also to its management and the construction of our urban environment, if only through advertising.

We could have discussed each branch of science in the same way as we have discussed astronomy here. For each discipline we would find a range of different interactions with the society in which it flourished. For each we would be able to identify fundamental contributions to our knowledge, applications transforming our very existence, and also political considerations. Science stands at a crossroads between so many different avenues of society and its proponents have a major responsibility in determining how we will manage our human and natural environments.

Chapter 4
Science: Pleasure and Culture

Science and Harmony

The professional lives of astronomers may well have been dominated for centuries by the problem of making precise, repetitive, and routine observations for the measurement of space and time, but they were always lived out under the night sky. Whatever the problems of their telescopes and instruments, the cold, tiredness, and other complications of working by night, none can completely escape from the fascination of this silent and solitary nighttime vigil beside a telescope. It is only very recently that astronomers have been able to operate their equipment while comfortably seated in a well lit and well heated control room. Despite the often biting cold in the dead of night in some high altitude observatory, eyes and thoughts turn easily toward the sky. Many questions come to mind, from philosophical introspection to simply trying to understand what it is we are looking at. Never far away is the pleasure of finding a certain harmony in the world. From the questioning inspired by a beautiful night sky, there may come an accrual of human knowledge, and from there an addition to human culture.

What we have said here for the centuries long work of astronomers can be transposed to any form of scientific activity. The daily work of the scientist has always required accuracy, attention to detail, writing up of reports, corrections to computer programs, setting up and repairing equipment, preparing samples, requesting funds, repeating measurements, and so on. Of course, there is a lot of routine work, often slow progress, and sometimes setbacks. Lack of understanding by financial backers and administrators generates frustration and suffering in science, just as it does in other professions. But the endless quest for harmony and theoretical elegance underlies the scientist's work and gives meaning to it. Notwithstanding the many constraints on their daily lives, they derive enough pleasure to keep on going.

Curiosity, the drive to know and understand, the satisfaction involved in achieving something, the pleasure of discovery, and the desire to contribute to the

© Springer International Publishing AG 2017
T.J.-L. Courvoisier, *From Stars to States*, SpringerBriefs in Astronomy,
DOI 10.1007/978-3-319-59232-9_4

cultural development of our civilisation are reasons for the young and the less young to turn to science. I don't think I have ever had any colleagues who have completed a scientific investigation and not been more or less motivated by these things. The contribution of their research to the economy and other utilitarian aspects of the sciences only come in much later, if at all.

Science and Pleasure

The study of science, in any branch, is above all a pleasure, the pleasure of discovering some of the most wonderful features of the world. It is the fascination of exploring some facet of reality, whether it be the structure of matter, the evolution of the Universe, or the molecular construction and processes of life. There is also the satisfaction of inventing and building different items, whether they be material things or software for computers. Student days are a very special experience for a young scientist. The current state of knowledge is presented in a coherent way, introducing the new arrival to what we know today by building everything up from the basics. It is only later that the *impedimenta* of professional life can sometimes tarnish the sparkle and brilliance of the research experience.

The study and practice of science also includes the pleasure of understanding. The privileged moment that Swiss Germans call *Aha Erlebnis*, a fleeting instant during which the various aspects of a problem fall into place and the solution of the problem suddenly becomes obvious. The student, researcher, or engineer then understands how the apparently unconnected pieces of a puzzle fit together to reveal some harmony of nature.

Such pleasures do not come easily, however. An engagement with the so-called hard sciences requires a certain intellectual discipline and asceticism, and such demands are in perfect contrast with the ways of the consumer society, whose advertising is forever presenting us with the promise of facility and immediate satisfaction. As in other aspects of life, the difficulty involved in the process is without doubt part of the pleasure one can gain from the final achievement. Landing by helicopter on top of a mountain is as nothing compared with a long climb to reach the peak. Approaching a new world on the keel of one's own boat after a long crossing leaves a quite different feeling than doing the same trip by plane. Learning about some branch of science is all about climbing slowly but surely toward an area of knowledge in the company of one's peers.

Completing scientific studies means climbing aboard the world as we know and understand it today. There is no book or paper one could read in order to go further. The next page is blank, desperately blank, and writing the next chapter is the challenge for everyone who takes part in research. A challenge which must first be met with an idea.

From Idea to Knowledge

It is generally while walking, sleeping, talking, reading, or pursuing some otherwise ordinary everyday occupation that an idea can suddenly come to us, quite unexpectedly. There it is, out of the blue, making us say to ourselves: "Why did I never think of that?" The generation of an idea is not a process that obeys rational rules, and it occurs regardless of whatever activities we may be engaged in, or the instantaneous preoccupations of our brain.

While most astrophysicists think that a structure in the form of a disk is the only conceivable explanation for the properties of quasars, just to give one example, we may wonder if the emissions we observe might not come from collisions between clouds of matter. This idea may take shape from a set of other ideas we carry around more or less consciously in our intellectual toolkit and pop out unexpectedly at some moment in our existence. And it may be the beginning of a process that could lead to the acquisition of a new piece of knowledge.

Once the idea has been formulated, it must be confronted with new observations that could either corroborate or falsify it. An idea leading to no measurable or observable consequence can be dropped immediately, for it will never bear fruit, and it will never lead to further knowledge. It falls outside the realm of reality. In most areas of science, the researcher will carry out an experiment that is specifically designed to test his or her idea. In astronomy, the scientist cannot design experiments in which parameters are controlled to determine causal relations. We can only observe phenomena that nature has chosen to present and attempt to decipher the message addressed to us by these observations. In the example above, a series of measurements on a quasar may allow us to measure the time dependence of its light intensity, something which can then be compared with the predictions following from our idea.

In order to carry out an experiment or a campaign of observations in astronomy, the means required may be relatively simple, but in many cases they will lie at the forefront of state-of-the-art technology. In the latter case, one must hire the skills of colleagues, technicians, or engineers, obtain financial backing, convince one's peers that the whole thing is a legitimate exercise in order to get access to suitable infrastructures, in our example the appropriate telescope, or again, design and build instruments that could lead to answers to the new questions. As scientists, we often spend the greater part of our time engaged in these long and arduous processes, far removed from the simple pleasures of understanding, and it is here that our work can become stressful at times.

Once measurements have been made, they must be interpreted and assessed in relation to the initial idea. But this is not all. Let us imagine that the idea is corroborated by the measurements or observations. It must still be fitted in with other knowledge relating to the same phenomena. So is our idea about quasars compatible with all the other things we know about these objects, and in particular the idea that the radiated energy comes from matter falling onto a black hole? It is only when our hypothesis is integrated into this broader context that it may be adopted into the general body of scientific knowledge. This process involves subjecting the idea and the measurements to criticism from colleagues with their own knowledge and experience of these matters, and not just in one's own department, but also elsewhere. Such confrontations are often intense, and the stakes are proportional to the effort invested to develop one's point of view. But it is essential if the idea is to grow from an individual preference and eventually become another building block in our universal knowledge.

Scientific theories are developed in the context of the day and it is only with great difficulty that new knowledge could ever completely overthrow this framework. A striking example of this phenomenon is Einstein's idea of including a cosmological constant in the equations of general relativity in order to incorporate a deeply rooted idea at the time, namely, the immutability of the Universe. But the story of the cosmological constant does not end with the discovery of the expansion of the Universe. Indeed, it made a spectacular comeback in the 1990s, when observations showed that the expansion of the Universe is in fact accelerating, whereas general relativity without a cosmological constant predicts that this expansion must gradually slow down. The exact nature of Einstein's term, which can be considered as a form of matter and which in that case is called dark energy, still defies our understanding. In this way, from the intellectual difficulties generated by confrontation between the consequences of a new theory and a deeply held collective belief was born a richer theory which only came into its own much later. The progress of scientific knowledge is no more a linear process than any other human activity.

Often enough the process is not as simple as the one described here. Observations or experiments will suggest modifications of the initial idea. They may also bring out completely unexpected effects. We then talk of discovery. The confrontation with already acquired knowledge or with work done by colleagues working in the same area can lead to incorporation of our measurements in a very different framework than the one initially intended. Above all, the process goes step by step, with new measurements leading to new ideas and so on.

It is the combined result of all these measurements, debate, criticism, confrontation with peers, integration with existing theories, and consequences in other contexts which makes the difference between an intuition or belief on the one hand and a piece of reliable knowledge on the other. We see and hear many assertions that are neither measured nor consistent with any body of reliable knowledge, but which are nevertheless presented as truths without ever being subjected to criticism by scientists. These assertions should be constantly countered and their authors and propagators sent before their peers, their test tubes, or their telescopes. It is easy to make such assertions in a forceful tone of voice, but a quite different matter to describe the idea and the process which could bring it into the fold of universal knowledge.

Science Can Be Communicated

The very rationalisation that underlies scientific knowledge makes the communication of scientific ideas possible. It makes all the difference with assertions beginning "I know" and "I believe". Any piece of knowledge can be presented in a reasoned way because it is connected to all the rest of human knowledge by an underlying rational process. Beliefs are barely open to any form of discussion beyond simply stating what is believed. There is no way of arguing the validity of a belief. It is up to each individual to believe it or not.

Once in place, an established piece of knowledge will serve as a foundation for further construction. New ideas and further reflection will be based on this new element, but may at any moment bring it back into question. It is only little by little that this collective construction can assume a stable form, without ever being completely safe from some deep reform, as happened with classical mechanics when relativity theory and quantum theory came on the scene at the beginning of the twentieth century.

The corpus of scientific knowledge can be understood by all, at least in principle. Whatever the starting point, whatever culture one belongs to, whatever knowledge is brought to bear, each and everyone can understand and appreciate the result of the work accomplished by successive generations of scientists on a given question. The unification of the various forces of nature was a combined effort that took more than two centuries. Among others, Michael Faraday set up experiments to understand electrical and magnetic phenomena, and he began to formulate the laws in the nineteenth century. His results were set in the form of four equations by James Clerk Maxwell later in the same century. Maxwell's equations unified the theories of electricity and magnetism with a rare elegance. This process was successfully pursued in the following century when electromagnetism was unified with the weak interactions and nuclear forces. And this process is ongoing as we attempt today to get all the forces of nature into the same conceptual framework. A considerable amount of work may be required to follow this line of inquiry or any other in the sciences, but the obstacles can always be overcome by investing the necessary time and effort. These considerations are based on the rationality of science, since it is founded on logic, and the axioms of logic are simple and known to all, at least intuitively.

Science as a Classical Aesthetic

Science is accessible to all and can always be debated. These are without doubt two of its strongest points. The price to pay for this achievement is a certain simplicity in the construction. Of all the ideas and concepts, the only ones to remain are those that can stand up to examination. The aesthetics of the resulting construction are thus closer to classical simplicity than to baroque extravagance. There are of course sticking points, features that look very much as though they don't belong, or are not really necessary for the construction. This is often the sign that new elements must be added to those already in place so that the whole thing can become consistent once more.

Paradoxically, one of the main difficulties in particle physics in the twenty-first century lies in the fact that the theory seems to be complete and provides a good explanation for the results of experiments in accelerators like the ones at CERN. This physics gives a good account of the matter we know, including protons, electrons, and other elementary particles, whereas we have learnt from astrophysical observations that 95% of the matter in the Universe actually escapes this description. The existing theory has few outstanding difficulties that might lead us toward a new model that could also encompass the exotic forms of matter which astronomy has revealed to us and which we have not yet been able to identify in our laboratories.

This monument we call scientific knowledge is characterised by classical simplicity rather than the frills and ostentation of the baroque. But classicism is not devoid of harmony or aesthetics. Elegance is a characteristic feature of any theory giving a satisfactory description of some aspect of reality. Some will go as far as to say that there is no "true" theory that is not elegant. Elegance, aesthetics, and harmony are profoundly human values, although for all we know irrational. However, the world as described by our science exists quite independently of us. By relating these two kinds of notion, we obtain a striking illustration of the fact that we ourselves belong to the physical world. This is probably what the author of Genesis wanted to say when asserting that man was made in the image of the Creator.

Science as a Source of Collective Culture

The observation of nature—taken in the broad sense as including humans and society—and the knowledge we can obtain from this, in a word, science, are closely related to various human cultures. Philosophy would be impossible without thinking about the place of the human being in the world, something which can be deduced from observations in astronomy, physics, biology, psychology, and sociology, and interpretation of these observations.

An important part of philosophy seeks to integrate scientific discovery into our thought systems. An example of such an endeavour can be found in the way Heisenberg's uncertainty principle has entered discussions about free will. This principle, corroborated by all kinds of experiments and observations, stipulates that we cannot simultaneously know both the position and the velocity of a particle with arbitrary accuracy. If the velocity is known, then the location of the particle must remain imprecise, and vice versa. It is thus impossible to predict the trajectory of a particle with total accuracy. This is a universal principle, hence valid for the electrons which determine our thoughts within the cells of our brains. This implies that there cannot be absolute determinism in the way our brains work. The implications of these considerations for human freedom have thus stood at the interface between science and philosophy for decades now.

The difficulty that certain societies and certain religions had to accept the Copernican system which places the Sun rather than the Earth, and hence humans, at the center of the Solar System illustrates the importance of the questions tackled by science for our systems of thought. The evolution of species discovered by biologists in the nineteenth century makes humans and their intelligence a mere link in a chain of evolution that will necessarily go on to new things. This is once again a contribution of key importance for philosophy and culture as a whole. The fact that certain groups or schools of thought have been unable to integrate what has by now become a fact of observation in their intellectual systems shows once again the importance of knowledge in the way we represent ourselves.

One consequence of the presence of knowledge within our thought systems is that religious systems can no longer remain completely closed, no matter what the wishes of those who lay down the ideology. Science will break down the door of belief systems that try to remain closed, basing themselves solely on the precepts of their respective authorities. Otherwise the price to pay is active rejection of ever more firmly established knowledge, as attested by fundamentalist Christians who have developed creationism or the theory of intelligent design. The refusal to perceive the evolution of the world around us, starting with the changes induced by volcanic activity, to give but one example, is a case of bad faith that cannot be attributed to ignorance alone. So knowledge is a natural tool for opening up intellectual systems, while attempts to keep such systems closed will involve ever greater effort and are doomed to failure.

Knowledge and Art

Art is not a science. The whole approach is different. But art, like all the rest of human thought, is nourished by knowledge. Painting techniques lie at the heart of the artistic research of the great painters. It is striking to read the letters that Vincent van Gogh wrote to his brother, Théo, where we find that he was constantly looking for new colours and new ways to apply them. When we look at his paintings for the first time, we are struck by their beauty and whatever it is we interpret as the message of the work, whereas the artist himself was very much focused on his technique. Likewise, architecture would be impossible without knowledge of physics, and sculpture without the knowledge involved in mastering the materials used.

Even literature maintains links with science. Knowledge contributes directly to literary texts. For example, Shakespeare used the knowledge of the day in many of his plays.[1] And astronomy has contributed to poetry, as observed by Jocelyn Bell Burnell (private communication), whose personal miscellany contains this sonnet by René François Sully Prudhomme[2]:

Le rendez-vous

Il est tard; l'astronome aux veilles obstinées,
Sur sa tour, dans le ciel où meurt le dernier bruit,
Cherche des îles d'or, et, le front dans la nuit,
Regarde à l'infini blanchir des matinées;

Les mondes fuient pareils à des graines vannées;
L'épais fourmillement des nébuleuses luit;
Mais, attentif à l'astre échevelé qu'il suit,
Il le somme, et lui dit: "Reviens dans mille années."

Et l'astre reviendra. D'un pas ni d'un instant
Il ne saurait frauder la science éternelle;
Des hommes passeront, l'humanité l'attend;

D'un œil changeant, mais sûr, elle fait sentinelle;
Et, fût-elle abolie au temps de son retour,
Seule, la Vérité veillerait sur la tour.

Encounter

It is late; astronomer in stubborn vigil,
From his tower, in the sky where the last sound dies,
Seeks golden islands, and, facing the night,
Endlessly contemplates the whitening mornings.

Worlds flee like winnowed grain;
Dense swarms of nebulas shine.
But attentive to some tousled heavenly body,
He calls to it, saying: "Return in a thousand years."

And it will return. Nor by a short step, nor by an instant,
Would it ever cheat eternal science;
Men will pass, but humanity will be expecting it;

With changing eye, but sure, like a sentry;
And if it be abolished by the time it return,
Truth alone will hold vigil over the tower.

---

[1]W.G. Guthrie, 1964, *Irish Astronomical Journal* Vol. 6 Number 6.
[2]René-François Sully Prudhomme, *Les épreuves*, 1866.

Or this verse by the American poetess Maura Stanton:

Computer map of the early universe
We're made of stars. The scientific team
Flashes a blue and green computer chart
Of the Universe across my TV screen
To prove its theory with a work of art:
Temperature shifts translated into waves
Of color, numbers hidden in smooth lines.
"At last we have a map of ancient Time"
One scientist says, lost in a rapt gaze.
I look at the bright model they've designed,
The Big Bang's fury frozen into laws,
Pleased to see it resembles a sonnet,
A little frame of images and rhyme
That tries to glitter brighter than its flaws
And trick the truth into its starry net.

Such examples clearly show that these are real contributions to literature and not just a verse version of what one might call "public understanding of science" or "popular science".

Theatre, film, or written word is almost always framed in a context given by nature as understood by the author or adapted by the author to the needs of the story to be told. If the scene is purely imaginary, it is still defined by some diverted version of nature. Whether it is a novel, an essay, a play, or a film, the work derives its meaning beyond this framework from the psychological study of a situation, often full of imagery. The author places his or her characters in a situation, often extraordinary, and then imagines what will happen. In doing so, he or she comes close to carrying out what Einstein called a *Gedankenexperiment*, an experiment that laboratory conditions rule out for technological reasons, but in which the laws of physics predict exactly what *should* happen. When Gustave Flaubert invented his character Madame Bovary, he was doing precisely this. By telling this story, he was analysing the reactions of Emma and her entourage as suggested by his under-standing of human nature. In this way, he saves us from having to actually live through the misfortunes of Madame Bovary and her husband, while nevertheless enriching our understanding of human behaviour.

Music is the harmony of sounds, and of numbers. In his book, *Gödel, Escher and Bach, an Eternal Golden Braid*, Douglas Hofstadter[3] brings together music, architecture, drawing, and mathematics in all their subtlety. He intertwines them into a braid just as each of his three heros wove together the elements of his art. This is a superb illustration of the way our thoughts can be nourished by mathe-matics, music, and the graphic arts.

---

[3]Basic Books, 1979.

Our civilisation is built for the main part on what we have learnt about nature over the millennia. Our philosophy, our approach to the world, and our art are profoundly influenced by what we know about the Earth, the Universe, or biology and human existence, in a word, by our scientific knowledge. In its turn, the society that has developed in this way guides the intellectual development of the researcher. The search for profound harmony in the world is the main concern of most of those who work to improve our knowledge.

Chapter 5
Knowledge, Management of the Environment, and Responsibility

The Impact of Fundamental Science on Society

Of all the sciences, astronomy is probably the one that seems to most people to be furthest removed from the practical concerns of society. And yet we have described here the many links between this knowledge and our everyday lives. These links extend from time measurement to applications of space science, geopolitics, technological developments, and industrial know-how.

The same can be said for particle physics. At the present time, the focus in this area is on the composition of matter and the forces acting on particles, at length scales so short that our own senses are of no use to comprehend them. While these investigations are well removed from our immediate economic and social concerns, it was physicists at CERN working in this field who instigated the development of the worldwide web, the archetypal technology when we think of things that have transformed the lives of just about everyone around the world.

These two sciences study phenomena at the largest and smallest scales in our Universe. They take us to the very limits of the Cosmos. Distances greater than the size of the observable Universe lie forever beyond our understanding and have no physical meaning, and likewise for those so small that the quantum uncertainty relations limit our access to them. But even these sciences of the extreme have led to important developments for society, not only in terms of pure knowledge and its influence on philosophy, but also in perfectly concrete and practical ways.

In many other areas, the link between the knowledge they generate and its impact on our lives is much more obvious. Thermodynamics, the study of heat and its relationship with other forms of energy, has been developed since the seventeenth century, when it was observed that cannons heat up when cannonballs are fired through them. These investigations then focused slowly but surely on the control of heat energy and its transformation into mechanical energy, the form of energy required for motion. This gave rise to the steam engine, used for example in

© Springer International Publishing AG 2017   39
T.J.-L. Courvoisier, *From Stars to States*, SpringerBriefs in Astronomy,
DOI 10.1007/978-3-319-59232-9_5

the nineteenth century to drive magnificent locomotives, then the development of the internal combustion engine, now used to power almost all our road vehicles.

Investigation of the nature of electricity and magnetism during the same period led to the theory of electrodynamics, a magnificent illustration of the power and elegance of mathematics when used to understand physical phenomena. But electrodynamics also underpins a whole series of important inventions: telecommunications, wireless or otherwise, are applications, along with electric motors, radar, and some of the technology in the computer on which these words have been inscribed.

Mathematics also belongs to this category of fundamental sciences that are rich in practical spinoffs. Its aim is to describe axiomatic systems of thought, the kind of a priori research that couldn't be further removed from everyday concerns. But we find its consequences in such down-to-earth considerations as the calculation of insurance premiums.

These sciences with their reputation for abstraction are thus much closer to the evolution and the important issues of our society than we might imagine.

The practical influence of the other natural sciences is often easier to recognise. Chemistry, biology, medical science, solid state physics, and the physics of other forms of matter have also contributed significantly to the evolution of our way of life. The materials around us, the electronics we use on a daily basis, and the medicines we take are recognisable and tangible consequences. In all the natural sciences there are fundamental and practical aspects.

Science and Innovation

When we consider more or less immediate applications of knowledge, often called innovation, fundamental science plays a dominant role. This kind of immediately applied science is based on a whole range of other knowledge that goes well beyond it. Materials science, which has been of enormous benefit to us over the decades and which we expect to deliver much more across many areas of technology, cannot be understood or taught without a deep understanding of quantum theory. To take a recent example, the foils used on boats to lift their hull out of the water cannot be designed without a deep understanding of fluid mechanics, which began with theories we owe to Bernouilli and Euler. It would also be delusory to think that students could be trained, even if it were just to play a role in the world of innovation, by teaching them solely about the most visible features of our understanding. It would be naive to think that we could innovate in anything but the most trivial way without rooting novelty in the rich earth of the underlying science. It thus turns out that the distinction between applied and fundamental science is often of little substance in practice.

But while all our knowledge may contribute to fashioning our culture and while all branches of science may influence the lives of each of us, we should not expect every study to lead to some direct practical application. The physics describing what goes on inside stars deals with the nuclear reactions that eventually allow them to shine, so there is a connection with the phenomenon of nuclear fusion, which some are hoping will provide a new way to meet our energy requirements here on Earth. On the other hand, there is little chance that the physics of matter accretion on black holes will ever contribute to solving the world's energy problems.

When we look back with hindsight at those things that have found practical applications, it is sometimes rather surprising. General relativity, Albert Einstein's great achievement, was published in 1915. His intention had been to bring intellectual consistency, on the one hand, to the theory of gravity which had been described by Newton, and on the other, to the observation that the speed of light is the same for all observers, whatever their relative velocity. Newtonian mechanics implies that information passes instantaneously between two gravitationally interacting bodies, while we know that information cannot be transmitted from one place to another faster than the speed of light. Removing this inconsistency was a purely intellectual research programme, and it was a success. However, the only observation he was then able to explain, and which had posed a problem for Newtonian mechanics, was a tiny rotation of the axis of the orbit of the planet Mercury, equal to just 43 s of arc per century. This is far from being a major consequence! However, a century later, relativistic corrections are needed to calculate positions using GPS to the required level of accuracy. An abstract exercise from which no one expected the least contribution to life in society has turned out to be one element of a tool that most of us now use on a daily basis.

In this example, not only does the relevant knowledge seem far removed from its application, but the time elapsed between the formulation of general relativity and its application is something like seven decades. When we reflect on science and society we must think not only of short term consequences, the kind of considerations dear to economists, but also much longer time scales than the length of a human life.

The Practical Scope of Human and Social Sciences

The human sciences are often criticised as moving in somewhat airy circles, disconnected from reality. Psychologists probe our dreams and our souls. But there too, acquired knowledge is often unashamedly used in the aggressive marketing that bombards us day after day. History, the study of a past that we are unable to influence, plays, or should play, a key part in the analysis of present situations carried out by our elected representatives, and it should help us to find rational solutions to conflicts of all kinds. Literature, to give one more example, describes behaviour in imagined situations and thereby extends our frame of reference in a very concrete way.

In other areas, knowledge is much less certain. For example, economic models disagree with one another and we struggle to deduce them rigorously from fundamental principles. Descriptions of different societies and political systems depend greatly on their authors. But this kind of knowledge, no matter how uncertain it may be, is nevertheless indispensable to the evolution of our society.

Knowledge and Responsibility

All our knowledge induces changes in the way we think, act, and fashion society. Every part of it can be viewed as a tool for controlling our physical and human environment. But whoever acts on, and has an effect on, their physical and human surroundings carries responsibility for this action. This responsibility is easily identified, even though it may be difficult to pin it down from the legal standpoint, when it comes to measuring the concrete effects of some technology. For example, internal combustion engines and heating using fossil fuels are technologies that increase the level of CO_2 in the atmosphere. We know who uses these techniques and who rejects what amounts of CO_2 into the atmosphere. We can thus identify the main parties responsible for this emission, and we know their contribution to the resulting climate change. The chain of responsibilities between the cause, the use of fossil fuels, and the effect, the change in the chemical composition of the atmosphere, is clearly established. Responsibilities with regard to the use of natural resources are also relatively easy to establish. We know who fishes and who eats fish, and we know the effects of this exploitation on ocean populations. The fruits of innovation, also a form of responsibility, are protected by patents designed to reward those who have provided their knowledge and know-how for some development. The author of a patent is identified so that she or he can profit from the economic consequences of the invention. The chain of responsibility is made explicit.

It is more difficult to define the responsibility of those who have contributed knowledge less directly connected with a given development. This kind of knowledge is usually in the public domain and it will have been published in the scientific literature. It may sometimes be rather old by the time of the application, as in the use of general relativity for GPS. The lapse of time between knowledge being acquired and its being made to act on the world, or the totally unpredictable nature of the consequences of some piece of knowledge, mean that the responsibility of the scientists whose results have contributed to an application may be rather difficult to determine. It would be hard to attribute to Einstein any responsibility in the development of satellite positioning systems! The chain of responsibility between a piece of research and its effects is then a rather poor way to describe the relationship between a researcher and society. However, researchers, the scholars of days gone by, bear a particular relation to a piece of knowledge through their occupation and perhaps also through their way of thinking, even for one that does not result directly from their own work. This relationship plays a role in society, involving once again a responsibility, independently of the tenuous connection that may exist between a given piece of research and an application.

Scientists, as researchers and teachers, have climbed the steps of what is known and crossed the boundaries between the known and the unknown precisely in order to shift them. They have shared their research with their peers, students, and the public. They have judged the knowledge of colleagues or students, and assessed the research of contemporaries. In short, they have spent hours and years to work and shape this understanding. This intimate association with science gives the researcher a special place in the relationship between knowledge and society. Nobody is better placed to gauge the relevance of work in progress, its value, and its weight relative to other work. This experience gives, or should give, a wider view of the relationship between knowledge and society, a view that goes well beyond the field of activity of any individual. So the idea is no longer to limit one's action to the focus of research of just one group, but rather to use the skills acquired to make science in general "genuinely useful for society", as specified in 1815 in the first statutes of the *Société Helvétique des Sciences Naturelles*, precursor of the present *Académie suisse des sciences naturelles*.

Like any other tool, those developed on the basis of established knowledge have an effect on the world. This is indeed what characterises them. But with such an influence comes responsibility for exploiting these capabilities. The problem here is not to determine who uses a technology stemming from some piece of knowledge, or with what aim. The point is that the practice of science and its results themselves have consequences for society. The acquisition of the tools that the proximity of this knowledge has given, or should give, then represents a responsibility that scientists must bear. This responsibility goes well beyond just communicating one's research results to the general public.

It is the duty of researchers financed by public funds to share acquired knowledge as widely as possible. This is done by publishing results in the scientific literature, taking part in conferences with one's peers, and organising communication, conferences, or articles for the general public. But although necessary, this form of communication is not enough to bring science into society in a useful way. A society in which science is only present through more or less effective conferences given by researchers to advertise their results, as and when they obtain them, will not know what to do with this knowledge. We still need to generate the culture required to put these new results in perspective, and above all to bring existing knowledge to places where it can and must contribute to actions fashioning the world. It must be allowed to bear fruit through suitable decision-making processes. This action is also a responsibility of the scientific community, as much as the direct consequences of its discoveries.

The gulf between science and its rational but complex processes on the one hand and decision-makers on the other requires those capable of establishing some kind of link between the two sides to actually get to work and do it. This task, which only scientists themselves can ensure, is part and parcel of their responsibilities with regard to the rest of society.

In an age sometimes referred to as the Anthropocene period to indicate that human action on the planet is now as important as geological phenomena, this responsibility is huge. It is knowledge that underpins humans' efforts to improve

their living space. It is thus also the source of both negative and positive consequences resulting from changes made to our local, and eventually also planetary,
environment. Researchers investigating the fundamental phenomena of nature bear
no greater direct responsibility for the possible effects of applying their discoveries
than the men and women who put this knowledge to use. Einstein has no
responsibility for consequences of the GPS. But this does not mean that the
researcher should not take full measure of the connection between human actions
and their consequences on the world. Indeed, this understanding is now crucial if
we are to keep our planet a hospitable place for its inhabitants.

What Knowledge Is at Work in the Evolution of Our Environment?

The complexity of the phenomena we have set in motion and the importance of
their consequences make it essential that modern scientists do everything in their
power to supply the public in general and political decision-makers in particular
with all the knowledge relevant to problem situations where action is possible or
necessary. And this body of knowledge is now enormous.

For example, consider the effects induced on the climate in which we all have to
live by changes in the composition of the atmosphere. This composition has always
changed over the ages. It has been influenced by biological activity, among other
things. An Earth without life would soon have an atmosphere with much less
oxygen in it, since this element bonds naturally with metals and would end up
binding to the Earth's surface and the ocean floor. But the change in the composition of the atmosphere observed at the moment is not caused by biological or
volcanic activity. For the main part it is due to the fact that, in just a few decades, by
burning fossil fuels like coal, we have been rejecting into the atmosphere certain
elements that biological and geological processes took millions of years to lock into
rock sediments. The speed of this change is without common measure with the time
required for these sediments to form. It means there is no way our environment will
be able to adapt to the new atmospheric conditions without some harmful consequences. We are already facing those consequences and must find some way to
limit them. Apprehending this reality, working out what exactly is involved, and
taking the necessary action constitutes a formidable challenge. We need to
understand very precisely how the atmosphere interacts with solar radiation, and
establish the physical and chemical interactions between the atmosphere and the
Earth's surface and the oceans. We must study the formation of clouds and their
optical properties on visible light and infrared radiation. We must monitor the
chemical reactions of all the atmospheric components and integrate them into our
calculations, but also the global circulation of the atmosphere and the oceans to
understand how heat is moved around at the Earth's surface. Each element of this
puzzle influences all the others, so they cannot be treated separately. We need to

study the effect each has on all the others. The challenge is enormous and the sum total of all the knowledge involved is commensurate with it.

Once the atmospheric, physical, and chemical phenomena have been understood, it still remains to work out what will be possible and necessary for the properties of the Earth's climate to evolve in such a way that the Earth remains a suitable place for the prosperity of all humans. This requires, and will continue to require, changes in our industrial practices and our way of life to halt the most harmful consequences of the transformations we are producing in the atmosphere. To take one example, we will certainly have to stop burning coal, despite the fact that this resource is still widely available around the world. But to achieve this, we need to understand the economic processes at work around the world, the possibilities for political action, and the reactions of individuals and the societies they belong to when such changes are made, and we will have to imagine new ways of producing and using heat energy. Taken together, a considerable amount of knowledge will have to be brought into this reflection from the human sciences, psychology, sociology, political science, economics, and technical sciences.

The world's population is growing steadily. This demographic problem forces us to think about food, among other things. Agricultural land is limited and the oceans are being heavily taxed. For the time being, ocean resources are mainly being exploited through fishing, in a way that often goes well beyond the capacity for fish stocks to regenerate naturally. We are behaving here like our hunter-gatherer ancestors, over-exploiting their territory. But our ancestors had the possibility of moving to new hunting grounds, whereas we have only the planet on which we live to feed our "tribe". A more intelligent use of sea resources would require us to understand the food chain in the oceans and develop some form of agriculture, that is, find ways to take into account photosynthesis by phytoplankton, the oceans' counterpart for plants on the land, and the biology of the animals that consume them, the aim being to develop a human food supply that will not exhaust the ecosystem. The knowledge required for such developments includes oceanography, but also biology,[1] tools for understanding the metabolism of the various species, and marine zoology and botany which are the macroscopic manifestations of these biological phenomena. This understanding is necessary, but it is still not sufficient for ocean resources to boost our food supply. To achieve this, we must accept to eat seafood. We shall need to understand nutritional culture and habits, hence culture in general, in different societies, and also how these cultures have evolved over time and how they might change again, in such a way that seafoods identified by new areas of investigation can become accepted. These changes must then be encouraged in some way, without this pushing people to reject them, without losing the richness of cultural diversity, and without those holding the knowledge arriving like conquerors to impose their own socioeconomic environment elsewhere. Once again, a whole range of knowledge must be brought together, just as much from the human sciences as from the natural sciences.

---

[1]*Marine sustainability in an age of changing oceans and seas*, EASAC report, 2016.

These are just two examples of the many challenges we must face up to. Among other areas we could have discussed are the problems of biodiversity, human and animal health, invasive species, and so on. They illustrate the body of knowledge humanity will need to build up to deal with the problems raised by a population of ten billion people who quite understandably would like to live a dignified and comfortable life.

It is the responsibility of the scientific community to acquire as much understanding as possible, then to communicate it to society, through politicians or the general public, in such a way that the decisions we will all be faced with on a global level are taken by people who are as well informed as possible.

Chapter 6
Knowledge and Politics

It is crucial to acquire the knowledge we need if society is to evolve in a good way. But knowledge alone cannot guarantee a better future for humanity. From it we still need to deduce what action is necessary and put it into practice. If knowledge is to bear fruit it must be brought to the political and economic scene, the spheres where decisions are made about the future of our society.

Knowledge obtained with the express goal of generating profit within a company will certainly be taken into consideration by the board of directors. Many large companies develop major programmes with this aim in mind. Quantitatively speaking, this kind of research is predominant in Switzerland: two thirds of research there is privately financed, for example, in the major pharmaceutical companies. Acquisition of this kind of knowledge is an integral part of the growth and development policy of such companies. Knowledge of a more general kind, as produced in the universities and other state research institutes, and publicly funded for the main part, is not destined to serve any particular industrial, institutional, or public policy. But can nevertheless be an important asset for the development of society. Its rather general nature, potentially useful in many different areas, which may lie far from any original intentions, make it more difficult for decision-makers to assess and assimilate than the results of research designed to achieve some specific aim.

Science Victorious

At the end of the Second World War, it was clear that science, and physics in particular, had played a major role in the victory over Nazism. Our understanding of the propagation of electromagnetic waves and its application in telecommunications

© Springer International Publishing AG 2017
T.J.-L. Courvoisier, *From Stars to States*, SpringerBriefs in Astronomy,
DOI 10.1007/978-3-319-59232-9_6

and the development of radar technology are two examples among many. In a famous report,[1] Vannevar Bush put forward the idea that scientific development could be beneficial, not only to the military, but also to civil society and the economy. He argued that human societies, and American society in particular, could also greatly benefit in an important way from science during peacetime. These ideas were widely approved and implemented during the ensuing decades. From the improvements observed in the standard of living in the West, it became clear that science was a boon to humanity. This attitude became deeply rooted in the Western world, to the point where the dialogue between society and the scientific community became considerably diminished during the 1980s, limiting itself to the public communication of research results. It was only later that a more critical view of science began to develop, and with it the need for renewed dialogue between science and society.

Political and Scientific Communities

The communication of scientific results has always been well received by an interested public. Sciences like astronomy still fascinate many. Open days in research centres regularly attract young and old alike. But it is rare to find actors on the political stage who are genuinely interested. Those involved in political life come more often from a legal, social, economic, agricultural, or purely political background than from any area of scientific activity, and in particular from areas involving the hard sciences like mathematics, physics, or chemistry. The skills required in careers that lead naturally to positions of political responsibility are essential to the work of any government and hence to the well-being of society. But it remains true that young people who follow this kind of syllabus often do so through disenchantment with science or even a certain aversion for it. Almost by design, therefore, most people involved in politics are barely familiar with the way science is done, the demands it makes, its power, and the results it can obtain.

Almost symmetrically, scientists rarely feel much affinity for the world of politics. The need to base one's knowledge and one's assertions on as many facts and experimental results as possible, the essence of the scientific approach, is hardly compatible with the current practice of political discourse to favour simplicity and immediacy. The distance maintained by scientists with regard to the political process, an attitude which in no way implies a lack of interest for social issues, often means that they are largely ignorant of the way decisions are made, and indeed the information needed to make them.

Moreover, the scientific and political communities approach questions very differently and they have diametrically opposite behavioural codes. The first pay

---

[1]"Science, the endless frontier", a report to the president on a program for postwar research by V. Bush, 1945.

little attention to form in the way they express themselves, whereas the latter consider them an important part of civility. These differences are reflected in the approach to dress. A photo of a group of research scientists at an international conference bears little resemblance to a similar picture of economic or political decision-makers. These differences add to the problem of communication between the two communities.

The lack of any particular need for such a dialogue between science and society for several decades, on the one hand, and the indifference or even fear of each side for the other, on the other hand, have allowed a barrier to build up that could now be difficult to break down. Their places of work and social activities are not the same. The two communities rarely meet. There are few personal connections between these two worlds and hence few opportunities for encounters that could build the necessary bridges.

Difficulties for the Dialogue Between Science and Politics

On top of the remoteness of the two communities, there are objective difficulties involved in transmitting scientific knowledge beyond the research community.

Research scientists are above all people moving at the frontier between what is known and what is not known, or not yet. Their whole approach is based on constantly questioning acquired knowledge until, little by little, their results find their place in the collective edifice of science. The researcher is therefore someone in a state of perpetual doubt, rather than someone inclined to categorical assertions. It is only when a piece of knowledge becomes relatively well established that scientists will assert their claims with assurance and demand sound arguments from those who would deny them. The knowledge required for a healthy development of society is often recent and complex, so scientific protagonists tend to be cautious in their claims and tend rather to preach doubt than certainty. This attitude is hard to reconcile with the politician's need to present easy-to-understand arguments in order to get some new policy adopted in society.

Scientific knowledge is built up on the measurement of experimental or observed quantities, such as series of temperature measurements in climatology. Measurements always have limited accuracy and thus involve some degree of uncertainty. Moreover, it is often necessary, not only to consider the past and present states of a system, but also to project into the future. This means that existing series of measurements must be extrapolated into the future, for example, to estimate the temperatures expected in some region of the world in coming decades. However, this extrapolation will depend on the models, which themselves involve inaccuracies, and these too add to the uncertainty in the results. The scientist will therefore indicate a margin of error describing the maximum uncertainty associated with each value. Once again the communication of such results outside the world of science is difficult and far removed from the need for simplicity felt by the public, its representatives, and decision-makers.

Research won't stop during this dialogue with society. It moves steadily on, while lawmakers work to integrate further knowledge into our social organisation. The transmission of knowledge can only follow its acquisition after some delay. The discovery of gravitational waves was only announced in February 2016, whereas it actually dates from September 2015. This delay was necessary to consolidate the result and make sure it was not just an artifact in the data. In a completely different context, our understanding of the impact that biofuel production may have on the production of food for people and animals has come a long way since the idea was first introduced, and has led to a reassessment of the advantages of this development, and hence also a reassessment of the way it should be carried out. Yet another difference adding to the difficulty of this dialogue.

The scientific community is often engaged in heated debate. Protagonists invest much time, energy, and effort to set up the experiments and observations they need, and then incorporate the resulting understanding into a world picture. They thus advocate their results before their peers with great forcefulness, confronting fellow scientists whose points of view may differ and who may present them with just as much enthusiasm. More time is then required, and further measurements, accompanied sometimes by a significant theoretical rethink, before the whole set of results can acquire the necessary level of consistency and serve as a basis for the next round of reflections or some concrete action in society. This culture of debate adds to the impression of uncertainty which surrounds newly acquired results, and to the difficulty in communicating those results.

In the discussions going on, an outside observer will struggle to distinguish those that deserve attention from those of no genuine relevance. For example, there are always some people who bring into question very well established facts, such as Einstein's special relativity or global warming, without providing any essentially new arguments. Such people can generate debates which find echo in the media or among the general public, and the latter, through lack of understanding of what is involved, may give equal weight to each side. These pointless controversies are then exploited by some to throw in doubt a whole field of established knowledge. In areas that are important for society, these discussions are used by people whose interests are affected by new knowledge to justify a resistance to change which is profitable to them. This is what has happened, for example, around the question of climate change. The facts are clearly established, but some are continually calling them into question, without bringing any new understanding to the matter, thereby sowing doubt in people's minds for reasons that have little to do with the common good. A useful scientific discussion must always be supported by observational or experimental facts, or by recourse to clearly identified theoretical elements. There should be no place for pointless infighting. It will then serve its purpose, and be worthy of the attention of its protagonists, the media, and the general public.

If the dialogue between science and society is to develop, all stakeholders must get together and reach an understanding. The world of science must be attentive to the needs of society. The questions society raises about research will be expressed in a form and in terms appropriate to the public and its representatives. However, these formulations will not always be in line with the scientific approach. For example, it makes sense to ask science to "cure cancer". And yet this is not a question that any researcher could tackle. This request must first be broken down into pieces that are well enough defined, hence small enough, to be dealt with separately with some degree of success. Each of these pieces will then become the target of a complete research programme, but the public will not always see the connection with the original question. Research that seems to be carried out well outside the context of the question can also make a relevant contribution to it, adding to the confusion. The translation of a request formulated by society into a scientific project is an important step in the dialogue. The scientific protagonists involved at this stage must listen carefully to the questions raised and make every effort to translate expectations into projects, in a way that is comprehensible to those who raised the question in the first place. Researchers may also have to elicit further questions on behalf of society, suggested either by specific research, or by the task of formulating other questions. And it should be remembered that the idea is not to "sell" all research as something that should necessarily contribute to solving the problems of the day. Indeed, when seeking funding for a given project, some groups will sing its praise in ways that go well beyond what can reasonably be expected of it.

Science and Power

Societal concerns often involve important political and economic questions. When we think for example of the development and control of seed production, for wheat, maize, or any other plant of importance for food supplies, those with the relevant know-how hold considerable economic power, not only the power associated with technological developments, but also the possibility of influencing the agri-food industry they serve. Political and economic players involved in running such a business would have a means of exerting a huge pressure to serve their own interests or ideology in countries that might lie well outside their jurisdiction. While biology offers particularly striking examples of the way knowledge can bring power, it is certainly not the only area where knowledge and power are bound so closely together. All the questions relating to energy and the climate also involve major economic and geopolitical stakes. And these are just a few examples among many. The dialogue between science and the world of politics is seriously complicated by the extent of the political and economic issues associated with so many areas of knowledge.

As already mentioned, an important part of knowledge is obtained through private research. This kind of knowledge is used primarily to develop companies and serve their interests, by which we mean the interests of their shareholders. Knowledge acquired in this context increases the economic power of the company and only becomes publicly available with its assent, hence only if a publication corresponds to its own objectives. The dialogue between science and society is thus partly biased, as certain decision-makers and scientists know things that are not accessible to either public or politicians. A parallel reflection is possible with regard to military research, often secret and designed to serve national interests. In reality, only results obtained in an open context, usually public, are available for reflection, decision-making, or action on the account of society as a whole.

The transition from open public research and industrial application, often essential if knowledge is to benefit society, receives a great deal of attention. It is hard to know which private bodies should take over from and pursue public research: when, how, with what level of risk, with what potential gain, and giving what role to public players? When research is applied, it is often this transition that determines the benefits that can be obtained, and who will be the beneficiaries. The latter will then hold the power from whatever knowledge and understanding is being applied.

There are also areas in which private research projects focus on themes with fast and tangible economic returns, leaving less immediately profitable themes to the public domain and state financing. This seems to be what happens in private pharmaceutical research, which is more interested in developing drugs for fighting the diseases of rich Western communities than those affecting the poorer regions of the globe. It is then the private sector which makes money and the public sector which takes on the challenge of solving questions that are sometimes of global importance, but without short or medium term gain.

Public Perception of Science

In addition to all these elements that obstruct straightforward communication between the world of knowledge and the decision-making bodies in our societies, there is a certain perception of science by a part of the general public which borders on the absurd. The measurement of climate parameters over very long periods of time and the inference that human action has a marked effect on the climate are some of the most solidly established facts we possess concerning our environment. And yet there is an important fringe of the population and its representatives who simply deny the evidence. On the other hand, when there is a rumour that the Earth is about to collide with an unknown planet, it will spread like wildfire, without a single observation, measurement, or other piece of evidence to back it up. There are so many people ready to believe such baloney without the slightest hesitation. Creationism is another invention that goes against all the observational evidence. Anyone seeing a stalactite in a cave understands that this cave will change over time. Anyone who has witnessed the changes wrought by a volcano will see that mountains are gradually transformed. And yet, there is a whole community that hangs on to beliefs in perfect contradiction with these observations. The efficiency of certain pseudo-medical practices such as homeopathy has never been established, despite endless studies, but that does nothing to prevent a large part of the population from adopting them, to the great advantage of the industries that promote them. These beliefs, which go against experiential fact, and the rejection of well-established evidence constitute yet another obstacle to the dialogue between the scientific community and society.

This dialogue is therefore a tough business, but that does not make it any less necessary if knowledge is to allow society to develop in harmony with its local and global environment. The first statutes of the *Académie suisse des sciences naturelles* in 1815—then called the *Société Helvétique des Sciences Naturelles* (SHSN)— expressed things like this: "The aim of the society is to encourage understanding of nature in general, and of the nature of our homeland in particular; to disseminate this knowledge and to apply it in a way that is genuinely useful for our homeland."[2] The notion of "society" has largely replaced that of "homeland" in everyday language. By replacing the word "homeland" by "society" in the last sentence of these statutes, we see that, even two centuries ago, they were already expressing the desire of people involved in science to use their knowledge in the service of the world around them. Our ancestors came remarkably close to the idea of *science for policy*.

Independence of Scientists in Their Role of Knowledge Provider

The concept known as "science for policy" should be clearly distinguished from "scientific policy". The first is about bringing knowledge into decision-making processes in all areas where it can or should be useful. The second, policy for science, is concerned with creating conditions in which teaching and research can flourish. In the latter, there is an aspect of working to favour the scientific community which is incompatible with the impartiality essential to the first.

We ask our political representatives and authorities to act for the common good of the whole collectivity. This same requirement carries over to those providing the knowledge that should underpin political action. In particular, they should not display any tendency, even indirect, to work in favour of their own community.

This independence must extend to any material, commercial, or financial interest. In order to be credible, a scientific opinion must be free, without censorship or self-censorship imposed by industrial interests, an employer, or indeed shareholders or other private or public lenders. It is generally the academies of science who take charge of providing scientific opinion. These organisations are the ones from which one would expect the highest level of independence. But it is an independence that must be zealously guarded. Academies or other institutions supplying knowledge to society must be explicitly free of any links with

---

[2]Statutes of the *Société helvétique des sciences naturelles* agreed in Berne on 3, 4, and 5 October 1816 and voted in Zurich on 6 October 1817; cited by Kupper and Schär, "Une organisation simple et sans prétention", in *Les Naturalistes, A la découverte de la Suisse et du monde 1800–2015*, Ed. Hier und Jetzt Baden 2015, p. 295.

commercial, financial, or industrial concerns, and free of religious, partisan, or political connections in the strict sense of the term.

Such independence could never be total. Some effort is needed to get a piece of knowledge into a form that can be used by society. For whatever subject, the corresponding data must be collected together, summarised, and expressed in a precise but comprehensible way, then transmitted. Each step here requires resources, time, and intelligence. And there is a price to pay, since scientists, like other members of society, must be able to lead a decent existence. Academies must therefore be financed in order to carry out their mission. The source of financing must be transparent and as far as possible free of conditions. It is usually the state that makes funds available for academies. Although this is the best way to guarantee independence, it is not completely free of constraints. An academy must sometimes draw attention to knowledge that contradicts a given state policy. For example, academies regularly support the idea that agriculture could become more efficient and more compatible with the environment by intelligently incorporating genetically modified organisms[3] (rather than doing this by successive selection), whereas the European Union and its member countries continue to proscribe the use of such methods. There is a tension here because lenders financing the academies must be prepared to fund work, even though the conclusions from this work are likely to support criticism of their position. In other words, financing made available to academies must be free of all influence. Such a demand is in contradiction with practices established over the past few decades, which are limited as far as possible to financing specific projects, chosen with the lender's agreement, of course.

Funding is not the only potential restriction on the independence of the academies or other organisations providing scientific opinion. The synthesis and presentation of knowledge are carried out by women and men. These individuals have their own life stories, with their own cultural and sometimes religious roots. The environment in which these scientists have forged their personalities will tend to influence their work and their vision. The very fact of devoting a part of one's life and one's time to science, then to take up the challenge of supplying knowledge to the decision-making process, implies a deep reflection on the meaning of human action and the value of knowledge. It would be impossible and quite probably not even desirable to ask scientists to disregard their cultural background and values in work they do to serve society. However, it is important as far as possible to be clear about the importance of such cultural components and to be able to spell them out whenever necessary.

Human relationships are also a form of dependence, connecting each of us to the rest of humanity. Scientists, like anyone else, love, dislike, and even hate some of their contemporaries. These relationships are an integral part of our personalities and are not without influence on the positions we adopt, even in areas like science and politics.

---

[3]See, for example, *Planting the future: opportunities and challenges for using crop genetic improvement technologies for sustainable agriculture*, EASAC policy report 21, June 2013.

Determining and declaring the allegiances of those involved in the dialogue between science and politics is a step of great importance for the credibility of a review or reflection.[4] In this respect, most academies have very explicit moral codes, at least with respect to commercial, financial, or industrial conflicts of interest. A second important measure if we are to ensure the validity of this kind of work is to set up groups of scientists from different social and cultural backgrounds, covering a broad spectrum of society. The responsibility of all the participants to examine their own prejudices and step back from them as far as possible is a third requirement.

A useful scientific discussion can usually separate observed fact and established knowledge from elements more closely tied to the cultural background of those involved in the discussion. The debate over genetically modified organisms provides a beautiful illustration of this tension. Such discussions take place in many European countries, in many different contexts, and within the scientific community itself. The relationship that each may have with nature, the leeway that each would allow to humans intervening directly on the evolution of organisms, in short the ethics of each individual, show through the opinions they express, while the experimental facts are clear and recognised by all honest and well informed individuals. A useful discussion is nevertheless possible and it can bring results provided that opinions and convictions are clearly recognised as such and distinguished from observational and experimental facts.

Naturally, the tools available to academies to guarantee the independence of those involved in science for policy from their counterparts in the dialogue between science and society generally ensure that no commercial or financial interest will taint the positions they adopt. However, these tools will always be limited to some extent when the problem is to assess the cultural and human aspects of their work. The responsibility of each person involved in the process remains the best guarantee of the quality of their work.

Scientists in the World of Politics

A very different form of dialogue between science and politics from the one proposed by academies and other institutions can be obtained by the direct participation of scientists in the world of politics. Around the world, there are scientists in research and teaching who play an active role in political life, in the narrow sense, by getting themselves elected to parliament, or even executive functions. However, these are exceptional cases. Politicians are often asked to speak about themes with which they have little familiarity, with the aim of making simplistic, even

---

[4]The author was educated in a strong Calvinist tradition. He studied theoretical physics and contributed to astrophysical research for more than three decades. He is active in the world of academies of science in Switzerland and Europe. So many elements that will influence what you are reading right now.

categorical assertions, and this is very different from scientific communication, which tends to be complex and which must constantly spell out the limitations of any given results. The transition from one way of working to the other is clearly difficult and can only be achieved by a few individuals. Of course, this state of affairs is regrettable, but the difference between these two attitudes means that the scientific component of a population will always be seriously under-represented in its political bodies. But then, without representation in parties and parliaments, scientists will make little contribution to societal debate.

Moreover, scientists taking part in the dialogue between knowledge and society are compelled to remain outside the debates leading to actual decisions, because they have to be independent of political parties, providing their knowledge to everyone regardless of party allegiances. It is thus paradoxical that the role of those who generate the knowledge underlying societal change, or who have produced the necessary synopsis so that it can be used in debate, is often limited to placing a report on the table and answering the occasional question about it.

This moral limitation which scientists accept voluntarily to ensure as far as possible their own independence is not shared by many non-governmental organisations who instrumentalise knowledge to serve their own purposes. These organisations are often very outspoken and active during debates. We may cite Greenpeace whose approach is perfectly legitimate, outside certain over-reactions, and whose voice expresses important positions of society, although sometimes rather too forcefully. It is nevertheless striking to observe the role played in 2014–15 by some of these organisations to counter the idea of a direct link between the authorities of the European Commission and the scientific community. The activism of certain non-governmental organisations should not be confused with efforts by the scientific community to make knowledge available to all without discrimination. The powerful voice based on strong convictions expressed by certain associations can sometimes smother even further the message that scientists could or should make heard in the general debate.

Despite the problems listed here, the dialogue between the worlds of science and politics has been renewed since the 1990s. In Switzerland, a popular vote destined to forbid the use of genetic engineering in 1998 functioned as a detonator. During the years and months prior to the vote, researchers realised that a gulf had opened up between the general population and themselves. They understood how important it was to build up a healthy relationship between society and their own work, not only for themselves, but also for society in general. They became more present on the communication platforms available to them. Many academies, including the *Académie suisse des sciences naturelles*, made major changes to their organisation so that they could play a bigger role in the life of society and work to establish a fruitful dialogue with their political counterparts.

New institutions were set up, like the European Academies Science Advisory Council (EASAC), whose aim is to bring knowledge and expertise to the institutions of the European Union. It remains to see to what extent these efforts will achieve the aim of making our planet a hospitable environment for all.

Chapter 7
Beyond Nations

From the Science of the Day to Universality

The scientist's work is rooted in a certain period of time, a certain place, and a certain culture. Which subjects are investigated in science, in philosophy, or simply down at the local pub, and indeed which are not, will vary from one region of the world to another and from one era to another. Geometry and cosmogony were subjects tackled by the Ancient Greeks, then taken up by the Arabs, before returning to the West in the Renaissance. The fields in fashion change with time even now, as they always have. The questions that interest one generation will not necessarily interest those of the next. A few decades ago, we smiled condescendingly at our colleagues studying asteroids in the Solar System, while this research has become one of great importance today. So research does not completely escape from social contingencies. As we have seen, its protagonists are fully fledged members of their community, and their work is inevitably influenced by their social, political, and cultural environment.

It is the huge effort of observation and experiment, accompanied by the confrontation between scientists and interpretations of measurements, which transforms research that may be infused with the cultural background of its authors into universally recognised results, independent of any geographical location or cultural consideration. Knowledge acquired like this is in principle accessible to anyone, anywhere around the world. It remains valid well beyond the time it was first established. In fact, it remains valid in the theoretical framework for which it was conceived. Newtonian mechanics gives an excellent description of gravity, provided that the speeds involved are negligible compared to the speed of light. General relativity takes over when this condition is no longer fulfilled. But general relativity has not replaced Newtonian mechanics. It has extended the area in which we now have an excellent description of the effects of gravity. We know that general relativity does not give a complete description of gravity, because it is not compatible with quantum theory. This restricts its field of application, but not its

T.J.-L. Courvoisier, *From Stars to States*, SpringerBriefs in Astronomy,
DOI 10.1007/978-3-319-59232-9_7

universality within this field of application. A new description of gravity must be found to supplement the one we use today. It must be compatible with general relativity in its area of applicability, that is, when quantum effects can be neglected.

Temporal, national, religious, political, and cultural boundaries are irrelevant when we try to understand scientific results. The measurement of climate parameters, including all their regional variations, cannot be taken as "correct" here and "false" somewhere else. They must be accepted everywhere, just like the theoretical models that are deduced from these measurements, once they have been rigorously tested. The ensuing discussion about what political action will be necessary to face up to this phenomenon must be based on these considerations. The action itself will depend on the political, economic, and social environment in a given region, and on its populations. But the effects of this action will once again have to be assessed scientifically, hence accepted without national or cultural prejudice.

The Earth as a Spacecraft

National boundaries have no meaning for science, and the same can be said when we consider the Earth as a single ecosystem. If we consider winds, ocean currents, animal migrations, and the movements of people and the various living species they carry with them in their luggage from one corner of the world to the other, all the evidence suggests that we should adopt a global view of our planet and remove the blinkers provided by political borders (Figs. 7.1 and 7.2).

Many actions undertaken in one part of the world, under the aegis of one nation, will have consequences in regions which may be geographically or temporally far

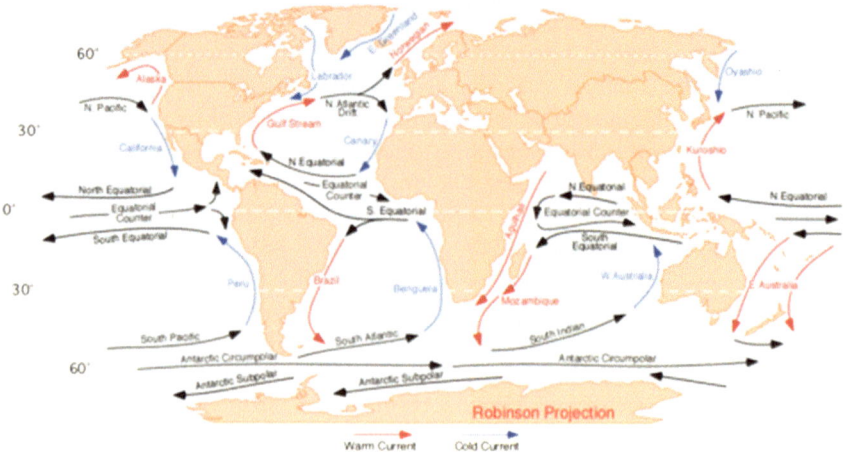

Fig. 7.1 Ocean currents illustrate the connections between the different oceans around the world. *Source* https://commons.wikimedia.org/wiki/File:Corrientes-oceanicas.png. *Credit* Dr. Michael Pidwirny

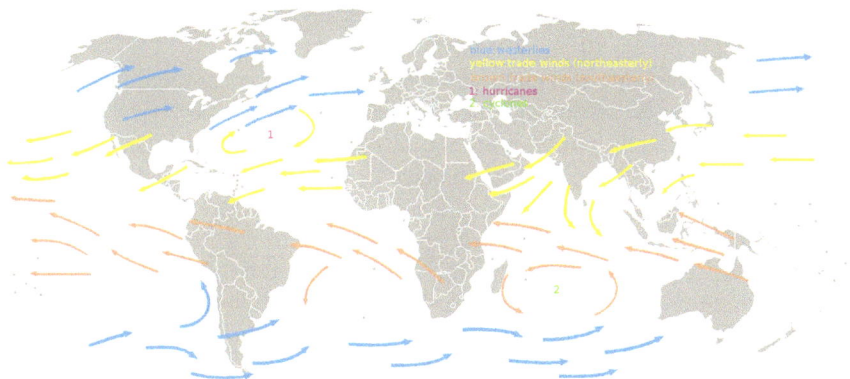

Fig. 7.2 Winds know no national boundaries. *Source* University of St Andrews

removed. Most of the chemical transformations in the atmosphere are consequences of Western industrial developments during the last century, but the effects of these transformations, such as the rise in sea level due to the oceans becoming warmer, are themselves influencing the whole planet and will do so for decades to come. Peoples living along the coast or on islands in the tropics are in no way responsible for the increased levels of greenhouse gases in the atmosphere, and yet the territories they depend on are being reduced and may disappear altogether. There are very probably examples of local actions whose distant or delayed effects are in fact beneficial. Indeed, one might consider a certain amount of warming of the coldest climates as being beneficial in some parts of the world and in certain circumstances. But notwithstanding, the global effects of industrialisation over past centuries, something we have now become fully aware of, are generally more problematic than anything else, and unlikely to bring any improvement to the human condition.

Decisions taken and the ensuing actions in one country or region are intended to have a beneficial effect on the local population or one of its sub-populations. The consequences of these actions on other populations, either close by or far away, are often ignored. Long term consequences, coming only after the mandates of those responsible for the decisions, are also often neglected, even when they are known. For this reason, many economies continue to orient their energy policies toward the consumption of fossil fuels, whereas in the interests of the planet as a whole, we should stop increasing the level of CO_2 in the atmosphere as soon as possible.

Power Is Concentrated in Countries

Our political decision-making systems, but also our economic ones to a large extent, have a distinctly national structure. It is on the scale of nations that most power is concentrated. Smaller subdivisions are always rooted in a national political system, and larger structures than nation states are few and far between and generally lack any operational power of their own.

However, nature knows no political boundaries and the consequences of some of our decisions may be relocated. The national context in which we fashion our evolution is thus seriously inadequate for the challenges raised by the influence of modern humanity on the global environment.

But there is nothing genetic about national borders. The extreme north of Europe illustrates the fact that certain forms of territorial organisation have existed up until now which are not based on the idea of a geographic nation. The Sami people who inhabited these regions lived alongside the seasonal movements of their reindeer herds until the end of the nineteenth century, when the Russians, Swedes, Finns, and Norwegians set up borders in this region. Each of these countries also established tax systems for those crossing the borders of their national territories, and this hampered such population movements, putting a financial load on their resources that was barely compatible with their income. The Sami lost out heavily because of these decisions, in which they played no part whatsoever. Their way of life was simply no longer viable, and indeed the construction of national borders in their living space proved fatal. The division of the territories of Africa and the Middle East into artificial "nations" by colonial powers was done without due regard for the local realities of those regions and provides another recent example of the way territories with a quite different organisation were compelled to become nations. In both cases, the construction of nation states has not been a great social success by any stretch of the imagination.

Even in the West in the twenty-first century, there are decision-making and operational structures which cross borders. These are the large multinationals and organised crime, entities which defy or even take advantage of national structures. Their economic and political success once again illustrates the limitations of the concept of nation state in the organisation of the modern world.

International Powers

The incompatibility between the demands made by the global nature of human action and the universality of science on the one hand and by an essentially national political structure on the other is regulated today by international treaties. These agreements result from negotiations between nations, sometimes under the aegis of international organisations. The relevant entities are predominantly states. They are founding members of the international organisations and signatories of the treaties. This is therefore always a matter of nations, a state of affairs represented even in the term "international" used to describe any action or organisation that is active on a wider scale than that of individual countries.

So the structure and composition of international organisations and also the elaboration of international treaties are the product of the balance of power between nations. The respective weights of the partners may come from the demographic importance, but also the economic, industrial, and military power of the stake-holders. But in the negotiations leading up to treaties in the framework of inter-national organisations, most parties will orient their demands and contributions toward defending the interests of the populations they represent, or should repre-sent, rather than considering the interests of the whole group. For example, the aim of the Science Programme Committee of the European Space Agency (the ESA, an intergovernmental organisation) in most of its activities at the beginning of the 2000s was to provide national scientific communities with the tools they wished to develop. The possibility of building a genuine European scientific space pro-gramme was eclipsed by the sum total of sometimes divergent interests of the ESA's member countries. But in a context set up for continental development, national interests are particular interests like any other and should rather be eclipsed by considerations of the general good.

The confrontation between divergent national interests sometimes leads, after endless disputes, to agreements that can bring progress on global environmental questions. A nice example is given by the problem of the stratospheric ozone layer. This molecule is important to protect life on Earth from the Sun's ultraviolet radiation, whether on land or on the ocean surface. The international scientific community became aware of the harmful effects of halogen gases on the density of ozone in the stratosphere and understood the connection between the depletion of these molecules and the presence of chlorofluorocarbons (CFC), a gas component used on an everyday basis and in particular for refrigeration. By 1987, once this had been established, the Vienna Convention and the Montreal protocol gave the necessary impetus to limit the emissions of CFCs that were causing depletion of the ozone layer.[1] We may now expect stratospheric ozone to persist in sufficient amounts to protect us from the Sun's harmful radiation. But it is no less true that weighing up national interests is certainly not the best way to arrive at a global environmental policy that can satisfy the whole of humanity. Examples abound to show that measures adopted in the available institutional framework, for example, to reduce greenhouse gas emissions, are not sufficient to control the chemical evolution of the atmosphere and its consequences for plant, animal, and human life.

Moves Toward Regional and Global Powers

Planet Earth is like a spacecraft moving through the empty space of the Solar System. Its only interaction with the surrounding Universe (almost[2]) is the Sun's radiation and the radiation that the Earth returns to space. For the rest, this vessel

---

[1]See, for example, Scientific Assessment of ozone depletion: 2010, World Meteorological Organisation Global Ozone Research and Monitoring Project, Report no. 52.

[2]Collisions with asteroids and cosmic rays coming from the Galaxy and beyond are exceptions. The influence of the Moon on the tides is another.

Fig. 7.3 The Earth setting as viewed from the Moon by the Japanese satellite Kaguya in 2007. 12:07 p.m. on November 7, 2007 (Japan Standard Time, JST). *Source* JAXA ©

must be self-sufficient, with no recourse to further intervention from anywhere beyond. It is thus essential that we manage it with this in mind (Fig. 7.3).

This responsibility has consequences for our decision-making system. In particular, it must be designed so that the measures agreed have influence only on the scale at which they are decided. Local decisions should have effects only locally, just as decisions made on a regional, national, continental, or global level should have effects only on the corresponding scale. It would be naive to think it would be easy to restrict the exercise of power to its legitimate field of influence, but if we could respect this rule in the establishment of the world order, it would be an important step toward a better management of the planet.

There is nothing simple about setting up a legitimate authority to make decisions in the name of the whole population of the world and endow it with the means necessary to ensure that those decisions are respected. For example, burning coal to heat one's house or produce electricity may well be locally advantageous, because coal is an abundant resource and relatively easy to extract in certain regions of the world, but it is one of the most serious sources of atmospheric pollution. It should be possible to take this effect into account and impose a drastic global reduction in coal consumption. But only a legitimate authority on the world stage could propose and endorse such a measure.

Implementing this vision thus means attributing real power to authorities acting above and beyond the national level. This happens in subdivisions of certain countries which have strong federal structures, endowed with well defined areas of authority and their own financial resources. The cantons of Switzerland and the communes they contain function in this way. Strongly centralised states like France are much less familiar with this form of power distribution. On a transnational scale, this vision requires a non-negligible part of national power to be put in the hands of continental and global organisations which are not just a representation of nations, but entities carrying responsibility for their action on their own scale and before all peoples who may be affected by their action. These authorities must have access to independent means so that they can analyse their own needs, take appropriate decisions, and ensure that they are respected. Among other things, this implies a source of funding that is independent of individual countries, so they must be able to raise their own taxes in some form or other.

The European Model

Europe is a continent on which this process has begun. Member states have given up some of their prerogatives to the benefit of the European Union, which is run by an autonomous parliament, the European Commission, and the Council of States. The parliament and the commission are purely European institutions, while the council represents member states. This means that the countries of Europe have kept a strong influence on the evolution of the European Union, but they don't hold all the power. This evolution has been painful, because national political bodies have found it difficult to delegate any of their authority, and only grudgingly hand over the power they have today to deal with problems of Europe-wide importance.

Progress toward a Europe that is responsible for its own future will require the constitution of a continental identity. For each citizen of the continent, being European should become as important as being Swiss, German, or Spanish. This is crucial if Europeans are to recognise themselves as stakeholders in the development of their own continent. But the process is difficult, because national identities have often been exacerbated over the centuries by the succession of military confrontations the continent has known. Federal countries are better prepared to follow such a path, since their inhabitants are used to having a regional identity as well as a national one. But even in Switzerland, a federation if ever there was one, in which cantonal identities are as strong as the Swiss identity itself, it remains extremely difficult to get the population to accept the need for and participate in a continental construction.

Just to make things even more difficult, certain populist political movements make their living by stirring up nationalistic feeling and using foreigners, even European ones, as scapegoats for all sorts of political and economic problems. They thus build up power within nations without regard for any common transnational cause. These movements find support from populations discouraged by the difficulties imposed by the modern world. They considerably increase the complexities involved in moving toward a transnational construction, despite the fact that this will be essential if we are to preserve a propitious environment for human development.

Despite all the difficulties encountered in the construction of Europe, in forums where Europeans work side by side with colleagues from other continents, it is striking to observe that the experience acquired in Europe is an important asset when setting up multinational collaborations. Europeans have an experience of cross-national discussion that few can match. It will thus be all the more difficult for other regions of the world to build on scales that are no longer merely national. And this all the more so in that certain continents are largely dominated by a single nation, such as the United States in North America. In these circumstances, building something new without always looking out for one's own interests will require a rare political skill.

The same thing will have to be achieved on the global level. There again, it will be crucial to endow some worldwide entity with a genuine decision-making and executive power that is sufficient to endorse its deliberations. As on the European level, this step will only be taken when each recognises his or her identity as an inhabitant of planet Earth, of the same importance as belonging to a continental, national, or regional community. This condition is indispensable if each human is to accept the need for global actions that are not just governed by a jumble of national considerations and national interests.

Scientific research does not entirely escape from competition between nations. But scientists nevertheless understand that the value of their results transcends any border. Over the decades, they have also become used to building major research tools in a continental or global context. Here the fact that Europe is made up of so many national units, each too small to compete on a global level, has given Europeans a head start in the development of transnational projects. But quite independently of this, the experience and desire of scientists to work together without geographical boundaries could be one of their main contributions to political construction on the continental or global scale.

Chapter 8
Science Is Not Everything in Life

Feelings

Our lives are built up from the opportunities that come our way and the decisions we take. Some of those decisions are inspired by our convictions and our feelings as much as by knowledge and reason. It would be absurd to ignore the role of emotion in our lives and choose our path through life in the light of knowledge and reason alone. For example, among the most important things in our lives are the development of romantic relationships, sometimes associated with raising a family. In the West, at the end of the twentieth century and the beginning of the twenty-first, the decisions we make over these matters are largely guided by our feelings and almost never involve social or economic issues. Reason can play a role here, to the extent that we adapt our desire to raise a family to the conditions society makes available. For example, this may concern arrangements to reconcile the various aspects of our professional and family lives. A knowledge of the habits and customs of society constitutes another rational element that some will take into account. Our family choices may also be influenced by demographic considerations on a regional or worldwide scale. But most of these decisions will be based rather on our feelings. Attempts to rationalise the search for a life partner using the algorithms of certain dating websites or, more traditionally, so-called marriage agencies, often seem somewhat ineffectual. Quite clearly, it is impossible to characterise people objectively in the hope of replacing an emotional approach to intimate human relations by an approach based on objective knowledge.

Sentimental relations are not the only aspects of our lives dominated by emotions. Many of the important acquisitions of our lives are made as much by our irrational attractions for a place, a style, or a trademark as by an anaysis of costs and benefits, whether it be a house, a boat, or a car. The world of advertising has certainly understood the emotional nature of these decisions and focuses more on exploiting such reactions than on informing us intelligently about the items or services we buy throughout our lives.

© Springer International Publishing AG 2017
T.J.-L. Courvoisier, *From Stars to States*, SpringerBriefs in Astronomy,
DOI 10.1007/978-3-319-59232-9_8

Our professional future is another area where reason and feelings fight it out in our brains to determine the path we will take through life. Reason would have us take into account the professional prospects of the various possibilities we consider, but emotional desires may take us along very different paths. It was doubtless unreasonable to study theoretical physics in the 1970s, but the way from there to writing these lines has been rich and fulfilling!

While love, friendship, or empathy are constructive feelings which allow us to embellish our lives and contribute to fulfilment in the lives of those close to us, jealousy, hatred, or indifference can harm or even destroy our social environment. Emotions are a driving force of human life, for better or for worse.

Collective Emotions and Convictions

Our lives together in society also have their emotions. Public interest in sporting events and the generosity of whole populations for certain fund-raising events is a visible expression of this. The reception and acceptance of Hungarian refugees in western Europe in 1956 or the wave of solidarity after the tsunami in the Indian Ocean in 2004 were beautiful examples of collective decisions which helped to reduce the suffering of whole populations. On the other hand, xenophobia and the rejection of others without trying to understand them have given rise throughout history to the worst atrocities. Here, too, emotions play important roles, whether for the common good or through the suffering they cause.

Convictions like emotions can be either constructive or destructive. They are no more rooted in knowledge and understanding than emotions. Religious faith is an example of a conviction that no amount of reasoning can found upon knowledge. It has motivated considerable and beautiful examples of generosity, but also some of the worst atrocities. The wars of religions which raged in Europe for centuries provide a striking illustration.

The Role of Emotions, Convictions, Knowledge, and Reason in Our Decisions

Little could be achieved if we did not take into account our emotions and convictions in our personal and collective decision-making processes. It would even be unreasonable not to do so, and it would diminish our faculty of discernment. But it is nevertheless important to use our knowledge and reason to recognise and analyse the role played in our behaviour by our feelings and thus give them their rightful place in our decisions. It is once again reason and knowledge that allow us to estimate the possible consequences of actions dictated by our feelings and understand their influence on our decision-making processes.

While the consequences of the decisions that each of us make over the years primarily affect ourselves and our own entourage, collective decisions have consequences going well beyond our own personal sphere of influence. In an age when human activities have such an effect on the environment in which we live that the well being, even the survival, of whole societies and populations is at stake, our decisions now carry an enormous weight. It is thus the responsibility of each authority to do everything possible to optimise the quality of its decisions. As well as mobilising all available knowledge, emotions must be gauged as objectively as possible, using reason to decide what weight to give them in the decisions we finally make. And finally, we must assess the religious or cultural convictions relevant to the problem at hand and, still as objectively as possible, analyse the consequences of the actions they inspire.

Knowledge must therefore find its place, not only in the objective appreciation of a situation, but also when we analyse the emotional aspects of that situation and assess the contribution of collective convictions to our reflections. All the tools made available to knowledge and reason to estimate the consequences of our decisions and our acts must be set in motion to ensure a good future for all the inhabitants of the Earth and their descendents.

We have an obligation to acquire the necessary knowledge without prejudice, weigh up particular interests, estimate the consequences of our acts and our decisions, and take into account our emotions and convictions with the necessary distance. History will judge our ability to act on the basis of all the available knowledge.